Specification lists

AQA B Mathematics

Refer to Revise AS for topics not listed below. Throughout the text all references to AQA refer to AQA B only.

MODULE	SPECIFICATION TOPIC	CHAPTER REFERENCE	STUDIED IN CLASS	REVISED	PRACTICE QUESTIONS
Pure 4 (P4)	*Algebra and functions*	*1.1; AS/2.1*			
	Coordinate geometry	*1.2*			
	Exponentials and logarithms	*1.3; AS/2.1*			
	Differentiation	*1.3; AS/1.5*			
	Integration	*1.4; AS/2.4*			
	Numerical methods	*1.5; AS/2.5*			
Pure 5 (P5)	*Coordinate geometry*	*1.2*			
	Sequences and series	*1.1*			
	Differentiation	*1.3; AS/1.5*			
	Integration	*1.4*			
	Numerical methods	*1.5; AS/2.5*			
	Vectors	*1.6; AS/3.1*			
Statistics 4 (S4)	*Continuous probability distributions*	*2.1*			
	Estimation	*2.3, 2.5*			
	Hypothesis testing	*2.4, 2.5, 2.6*			
Statistics 5 (S5)	*Normal distribution*	*2.3; AS/4.4*			
	Estimation	*2.3*			
	Hypothesis testing	*2.4*			
Mechanics 2 (M2)	*Newton's laws of motion*	*3.1; AS/3.4*			
	Application of differential equations	*3.1*			
	Kinematics	*3.1; AS/3.2*			
	Uniform circular motion	*3.6*			
	Work and energy	*3.7*			
	Centre of mass	*3.3, 4.5*			
	Simple harmonic motion	*4.3*			
Mechanics 3 (M3)	*Kinematics in two and three dimensions*	*4.1*			
	Uniform circular motion	*3.6, 4.4*			
	Work and energy	*3.7*			
Discrete 2 (D2)	*Recurrence relations*	*5.2*			
	Coding	*5.3*			

Examination analysis

The Advanced Level Mathematics GCE consists of AS (50%) and A2 (50%). AS consists of P1 + P2 + one of M1, S1, D1. A2 consists of P4 + P5 + one of S4/5, M2/M3 or D2 (following any dependency rules).

P4	**A2**	*May use graphics calculator*	*1 hr 15 min exam*	*15% of A Level*
P5	**A2**	*Scientific calculator only*	*1 hr 15 min exam*	*15% of A Level*
S4 or S5	**A2**	*May use graphics calculator*	*1hr 45 min exam*	*20% of A Level*
M2 or M3	**A2**	*May use graphics calculator*	*1hr 45 min exam*	*20% of A Level*
D2	**A2**	*May use graphics calculator*	*1 hr 15 min exam*	*20% of A Level*

Revise

A2

Mathematics

Peter Sherran
& Janet Crawshaw

Contents

Edexcel Mathematics

MODULE	SPECIFICATION TOPIC	CHAPTER REFERENCE	STUDIED IN CLASS	REVISED	PRACTICE QUESTIONS
Pure 3 (P3)	Algebra	1.1; AS/2.1			
	Coordinate geometry	1.2			
	Series	1.2			
	Differentiation	1.3; AS/2.3			
	Integration	1.4			
	Vectors	1.6; AS/3.1			
Statistics 2 (S2)	The binomial and Poisson distributions	2.2; AS/4.3			
	Continuous random variables	2.1			
	Continuous distributions	2.1, 2.2			
	Hypothesis tests	2.3, 2.4			
Statistics 3 (S3)	Combinations of random variables	2.1			
	Sampling	2.3			
	Estimation: confidence intervals and tests	2.3, 2.4			
	Goodness-of-fit and contingency tables	2.6			
	Regression and correlation	2.5; AS/4.5			
Mechanics 2 (M2)	Kinematics	3.2; AS/3.2			
	Centres of mass	AS/3.3			
	Work and energy	3.7			
	Collisions	3.5			
	Statics of rigid bodies	3.4; AS/3.3			
Mechanics 3 (M3)	Further kinematics	4.1			
	Elastic springs and strings	4.2			
	Further dynamics	4.1, 4.2			
	Motion in a circle	3.6, 4.4			
	Statics of rigid bodies	3.4, 4.5			
Decision 2 (D2)	Transportation and allocation problems	–			
	The travelling salesman problem	AS/5.2			
	Game theory	5.1			
	Dynamic programming	–			

Examination analysis

The Advanced Level Mathematics GCE consists of AS (50%) and A2 (50%). AS consists of P1 + P2 + one of M1, S1 or D1. A2 consists of P3 + two application units (following any dependency rules and including at least one A2 unit).

P3	A2	Scientific calculator only	1 hr 30 min exam	$16\frac{2}{3}$% of A Level
M2	A2	May use graphics calculator	1 hr 30 min exam	$16\frac{2}{3}$% of A Level
M3	A2	May use graphics calculator	1 hr 30 min exam	$16\frac{2}{3}$% of A Level
S2	A2	May use graphics calculator	1 hr 30 min exam	$16\frac{2}{3}$% of A Level
S3	A2	May use graphics calculator	1 hr 30 min exam	$16\frac{2}{3}$% of A Level
D2	A2	May use graphics calculator	1 hr 30 min exam	$16\frac{2}{3}$% of A Level

OCR Mathematics

MODULE	SPECIFICATION TOPIC	CHAPTER REFERENCE	STUDIED IN CLASS	REVISED	PRACTICE QUESTIONS
Pure 3 (P3)	Rational functions and binomial theorem	1.1			
	Coordinate geometry	1.2			
	Trigonometry	AS/2.2			
	Differentiation	1.3			
	Integration	1.4			
	First order differential equations	1.4			
	Vectors	1.6; AS/3.1			
Statistics 2 (S2)	Continuous random variables	2.1			
	The normal distribution	2.2; AS/4.4			
	The Poisson distribution	2.2; AS/4.3			
	Sampling and hypothesis tests	2.3, 2.4			
Statistics 3 (S3)	Continuous random variables	2.1			
	Linear combinations of random variables	2.1			
	Confidence intervals; the t-distribution	2.3, 2.5			
	Difference of population means and proportions	2.4, 2.5			
	χ^2 tests	2.6			
Mechanics 2 (M2)	Centre of mass	3.3			
	Equilibrium of a rigid body	3.4			
	Motion of a projectile	3.1			
	Uniform motion in a circle	3.6			
	Coefficient of restitution; impulse	3.5			
	Energy, work, power	3.7			
Mechanics 3 (M3)	Equilibrium of rigid bodies in contact	–			
	Elastic springs and strings	4.2			
	Impulse and momentum in two dimensions	4.6			
	Motion in a vertical circle	4.4			
	Linear motion under a variable force	4.1			
	Simple harmonic motion	4.3			
Discrete 2 (D2)	Game theory	5.1			
	Flows in a network	AS/5.2			
	Matching and allocation problems	AS/5.5			
	Critical path analysis	AS/5.3			
	Dynamic programming	–			

Examination analysis

The Advanced Level Mathematics GCE consists of AS (50%) and A2 (50%). AS consists of P1 + P2 + one of M1, S1 or D1. A2 consists of P3 + two other units (following any dependency rules and choosing not more than one from M1, S1, D1; not more than one from P4, C1 (Project); at least one from P4, M2, M3, S2, S3, D2).

P3	A2	Scientific calculator only	1 hr 20 min exam	$16\frac{2}{3}$% of A Level
M2	A2	May use graphics calculator	1 hr 20 min exam	$16\frac{2}{3}$% of A Level
M3	A2	May use graphics calculator	1 hr 20 min exam	$16\frac{2}{3}$% of A Level
S2	A2	May use graphics calculator	1 hr 20 min exam	$16\frac{2}{3}$% of A Level
S3	A2	May use graphics calculator	1 hr 20 min exam	$16\frac{2}{3}$% of A Level
D2	A2	May use graphics calculator	1 hr 20 min exam	$16\frac{2}{3}$% of A Level

WJEC Mathematics

MODULE	SPECIFICATION TOPIC	CHAPTER REFERENCE	STUDIED IN CLASS	REVISED	PRACTICE QUESTIONS
Pure 2 (P3)	Algebra	1.1			
	Coordinate geometry	1.2			
	Differentiation	1.3; AS/2.3			
	Integration	1.4; AS/2.4			
	Trigonometry	AS/2.2			
	Vectors	1.6			
Statistics 2 (S2)	Continuous distributions	2.1; AS/4.4			
	Combinations of random variables	2.1			
	Normal distributions	2.1; AS/4.4			
	Hypothesis testing	2.4			
Statistics 3 (S3)	Estimation	2.3			
	Confidence intervals	2.3			
	Hypothesis testing	2.4			
	Bivariate distributions	AS/4.5			
Mechanics 2 (M2)	Rectilinear motion	3.2			
	Dynamics of a particle	3.5, 3.7, 4.1, 4.2			
	Motion under gravity in two dimensions	3.1			
	Vectors in two and three dimensions	3.2, 3.5, 3.7; AS/3.1			
	First order differential equations	4.1			
Mechanics 3 (M3)	Second order differential equations	–			
	Rectilinear motion	4.1			
	Simple harmonic motion	4.3			
	Circular motion	3.6, 4.4			
	Statics	3.4			
Mechanics and Statistics (M5)	Continuous distributions	2.1; AS/4.4			
	Combinations of random variables	2.1			
	Normal distribution	2.2; AS/4.4,			
	Dynamics of a particle	3.7			
	Motion under gravity	3.1			
	Statics	3.4			

Examination analysis

The Advanced Level Mathematics GCE consists of AS (50%) and A2 (50%). AS consists of P1 + P2 + one of M1 or S1. A2 consists of P3 + two application units (following any dependency rules and including at least one A2 unit).

P3	A2	May use graphics calculator	1 hr 30 min exam	$16\frac{2}{3}$% of A Level
M2	A2	May use graphics calculator	1 hr 30 min exam	$16\frac{2}{3}$% of A Level
M3	A2	May use graphics calculator	1 hr 30 min exam	$16\frac{2}{3}$% of A Level
S2	A2	May use graphics calculator	1 hr 30 min exam	$16\frac{2}{3}$% of A Level
S3	A2	May use graphics calculator	1 hr 30 min exam	$16\frac{2}{3}$% of A Level
MS	A2	May use graphics calculator	1 hr 30 min exam	$16\frac{2}{3}$% of A Level

NICCEA Mathematics

MODULE	SPECIFICATION TOPIC	CHAPTER REFERENCE	STUDIED IN CLASS	REVISED	PRACTICE QUESTIONS
Pure 3 (P3)	Algebra and functions	1.1			
	Coordinate geometry	1.2			
	Series	1.1			
	Trigonometry	AS/2.2			
	Differentiation	1.3			
	Integration	1.4			
	Numerical methods	1.5			
	Vectors	1.6; AS/3.1			
Statistics 2 (S2)	Probability distributions	2.1			
	Statistical inference	2.3, 2.4			
	Correlation and regression	2.5; AS/4.5			
Statistics 3 (S3)	Further probability distributions	2.2			
	Further statistical inference	2.3, 2.4, 2.5, 2.6			
Mechanics 2 (M2)	Integration and differentiation of vectors	3.2			
	Variable acceleration	3.2			
	Uniform circular motion	3.6			
	Hooke's law	4.2			
	Further particle equilibrium	4.2			
	Work and energy	3.7			
	Power	3.7			
	Projectiles	3.1			
Mechanics 3 (M3)	Centre of mass of uniform laminae	3.3			
	Centre of mass of uniform solids	4.5			
	Force in two dimensions	3.4; AS/3.4			
	Direct impact	3.5			
	Resultant velocity	AS/3.1			
	Universal law of gravitation	–			
	Simple harmonic motion	4.3			

Examination analysis

The Advanced Level Mathematics GCE consists of AS (50%) and A2 (50%). AS consists of P1 + P2 + one of M1, S1 A2 consists of P3 + two of M1, M2, M3, S1, S2, S3 (following any dependency rules and containing at least one of S2 and M2).

P3	*A2*	*May use graphics calculator*	*1 hr 30 min exam*	*$16\frac{2}{3}$% of A Level*
M2	*A2*	*May use graphics calculator*	*1 hr 30 min exam*	*$16\frac{2}{3}$% of A Level*
M3	*A2*	*May use graphics calculator*	*1 hr 30 min exam*	*$16\frac{2}{3}$% of A Level*
S2	*A2*	*May use graphics calculator*	*1 hr 30 min exam*	*$16\frac{2}{3}$% of A Level*
S3	*A2*	*May use graphics calculator*	*1 hr 30 min exam*	*$16\frac{2}{3}$% of A Level*

AS/A2 Level Mathematics courses

AS and A2

All Mathematics A Level courses being studied from September 2000 are in two parts, with three separate modules in each part. Students first study the AS (Advanced Subsidiary) course. Some will then go on to study the second part of the A Level course, called A2. Advanced Subsidiary is assessed at the standard expected halfway through an A Level course: i.e., between GCSE and Advanced GCE. This means that the AS and A2 courses are designed so that difficulty steadily increases:

- AS Mathematics builds from GCSE Mathematics
- A2 Mathematics builds from AS Mathematics.

How will you be tested?

Assessment units

For AS Mathematics, you will be tested by three assessment units. For the full A Level in Mathematics, you will take a further three units. AS Mathematics forms 50% of the assessment weighting and A2 Mathematics forms the other 50% for the full A Level.

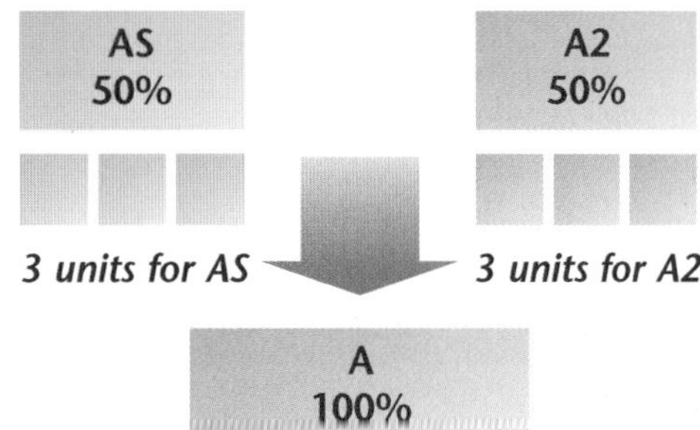

Most units can be taken in either January or June, but some A2 units must be taken at the end of the course. There is a lot of flexibility about when exams can be taken and the diagram below shows just some of the ways that the assessment units may be taken for AS and A Level Mathematics.

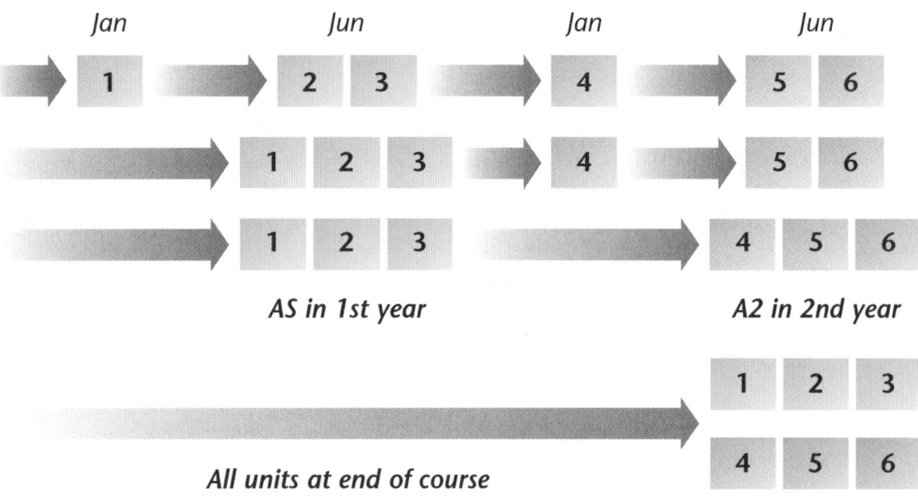

If you are disappointed with a module result, you can resit each module once. You will need to be very careful about when you take up a resit opportunity because you will have only one chance to improve your mark. The higher mark counts.

A2 and Synoptic assessment

After having studied AS Mathematics, to continue studying Mathematics to A Level you will need to take three further units of Mathematics at A2. Some A2 units draw together different parts of the course in a 'synoptic' assessment.

Coursework

Coursework may form part of your A Level Mathematics course, depending on which specification you study. Where students have to undertake coursework, it is usually for the assessment of practical skills but this is not always the case.

Key skills

It is important that you develop your key skills throughout your AS and A2 courses. These are important skills that you need whatever you do beyond AS and A Levels. To gain the key skills qualification, which is equivalent to an AS Level, you will need to collect evidence together in a 'portfolio' to show that you have attained Level 3 in Communication, Application of number and Information technology. You will also need to take a formal testing in each key skill. You will have many opportunities during AS and A2 Mathematics to develop your key skills.

It is a worthwhile qualification, as it demonstrates your ability to put your ideas across to other people, collect data and use up-to-date technology in your work.

What skills will I need?

For A Level Mathematics (AS and A2), you will be tested by assessment objectives: these are the skills and abilities that you should have acquired by studying the course. The assessment objectives for A Level Mathematics are shown below.

Candidates should be able to:

- recall, select and use their knowledge of mathematical facts, concepts and techniques in a variety of contexts
- construct rigorous mathematical arguments and proofs through use of precise statements, logical deduction and inference and by the manipulation of mathematical expressions, including the construction of extended arguments for handling substantial problems presented in unstructured form
- recall, select and use their knowledge of standard mathematical models to represent situations in the real world; recognise and understand given representations involving standard models; present and interpret results from such models in terms of the original situation, including discussion of the assumptions made and refinement of such models

- comprehend translations of common realistic contexts into Mathematics; use the results of calculations to make predictions, or comment on the context; and, where appropriate, read critically and comprehend longer mathematical arguments or examples of applications
- use contemporary calculator technology and other permitted resources (such as formulae booklets or statistical tables) accurately and efficiently; understand when not to use such technology, and its limitations; give answers to appropriate accuracy.

Exam technique

Prior level of attainment

Mathematics is, inherently, a sequential subject. There is a progression of material through all levels at which the subject is studied. The criteria therefore build on the knowledge, understanding and skills established at GCSE.

Thus, candidates embarking on AS/A2 Level study in Mathematics subjects are expected to have achieved at least Grade C in GCSE Mathematics, or equivalent, and to have covered all the material in the Intermediate Tier. In addition, candidates will be expected to be able to use the material listed below whenever it is required. This material, together with the GCSE material, is regarded as assumed background knowledge. However, it may be assessed within questions focused on other material from the relevant specification.

Background knowledge

For AS and A2 Mathematics

- The arithmetic of integers (including HCFs and LCMs), of fractions, and real numbers.
- The laws of indices for positive integer exponents.
- Solution of problems involving ratio and proportion (including similar triangles, and links between length, area and volume of similar figures).
- Elementary algebra (including multiplying out brackets, factorising quadratics with integer coefficients, to include $a^2 - b^2$, and solution of simultaneous linear equations by eliminating a variable).
- Changing the subject of a simple formula or equation.
- The equation $y = mx + c$ for a straight line; gradient and intercept.
- The distance between two points in 2-D with given coordinates.
- Solution of triangles using trigonometry, including the sine and cosine rules.
- Volume of cone and sphere.
- The following properties of a circle:
  (i) the angle in a semicircle is a right angle
  (ii) the perpendicular from the centre to a chord bisects the chord
  (iii) the perpendicularity of radius and tangent.

For A2 Mathematics

For most A2 modules you will be asked to demonstrate knowledge of a module already studied, for example the content of P1 and P2 is assumed for module P3. Check your specifications for these 'dependency' rules.

What are the examiners looking for?

Examiners use certain words in their instructions to let you know what they are expecting in your answer. Make sure that you know what they mean so that you can give the right response.

Write down, state

You can write your answer without having to show how it was obtained. There is nothing to prevent you doing some working if it helps you, but if you are doing a lot then you might have missed the point.

Calculate, find, determine, show, solve

Make sure that you show enough working to justify the final answer or conclusion. Marks will be available for showing a correct method.

Deduce, hence

This means that you are expected to use the given result to establish something new. You must show all of the steps in your working.

Draw

This is used to tell you to plot an accurate graph using graph paper. Take note of any instructions about the scale that must be used. You may need to read values from your graph.

Sketch

If the instruction is to sketch a graph then you don't need to plot the points but you will be expected to show its general shape and its relationship with the axes. Indicate the positions of any turning points and take particular care with any asymptotes.

Find the exact value

This instruction is usually given when the final answer involves an irrational value such as a logarithm, e, π or a surd. You will need to demonstrate that you can manipulate these quantities so don't just key everything into your calculator or you will lose marks.

If a question requires the final answer to be given to a specific level of accuracy then make sure that you do this or you might needlessly lose marks.

Some dos and don'ts

Dos

Do read the question

- Make sure that you are clear about what you are expected to do. Look for some structure in the question that may help you take the right approach.
- Read the question *again* after you have answered it as a quick check that your answer is in the expected form.

Do use diagrams

- In some questions, particularly in mechanics, a clearly labelled diagram is essential. Use a diagram whenever it may help you understand or represent the problem that you are trying to solve.

Do take care with notation

- Write clearly and use the notation accurately. Use brackets when they are required.
- Even if your final answer is wrong, you may earn some marks for a correct expression in your working.

Do learn relevant formulae

- Each module specification contains a list of formulae that *will not be given* in the examination. Make sure that you learn these formulae and practise using them.
- Some formulae *will be given* in the Examination Formulae Booklet. Make sure you know what they are and where to find them. This will help you refer to them quickly in the examination so that you don't waste time.

Do avoid silly answers

- Check that your final answer is sensible within the context of the question.

Do make good use of time

- Choose the order in which you answer the questions carefully. Do the ones that are easiest for you first.
- Set yourself a time limit for a question depending on the number of marks available.
- Be prepared to leave a difficult part of a question and return to it later if there is time.
- Towards the end of the exam make sure that you pick up all of the easy marks in any questions that you haven't got time to answer fully.

Don'ts

Don't work with rounded values

- There may be several stages in a solution that produce numerical values. Rounding errors from earlier stages may distort your final answer. One way to avoid this is to make use of your calculator memories to store values that you will need again.

Don't cross out work that may be partly correct

- It's tempting to cross out something that hasn't worked out as it should. Avoid this unless you have time to replace it with something better.

Don't write out the question

- This wastes time. The marks are for your solution!

What grade do you want?

Everyone should be able to improve their grades but you will only manage this with a lot of hard work and determination. The details given below describe a level of performance typical of candidates achieving grades A, C or E. You should find it useful to read and compare the expectations for the different levels and to give some thought to the areas where you need to improve most.

Grade A candidates

- Recall or recognise almost all the mathematical facts, concepts and techniques that are needed, and select appropriate ones to use in a variety on contexts.
- Manipulate mathematical expressions and use graphs, sketches and diagrams, all with high accuracy and skill.
- Use mathematical language correctly and proceed logically and rigorously through extended arguments or proofs.
- When confronted with unstructured problems they can often devise and implement an effective solution strategy.
- If errors are made in their calculations or logic, these are sometimes noticed and corrected.
- Recall or recognise almost all the standard models that are needed, and select appropriate ones to represent a wide variety of situations in the real world.
- Correctly refer results from calculations using the model to the original situation; they give sensible interpretations of their results in the context of the original realistic situation.
- Make intelligent comments on the modelling assumptions and possible refinements to the model.
- Comprehend or understand the meaning of almost all translations into mathematics of common realistic contexts.
- Correctly refer the results of calculations back to given context and usually make sensible comments or predictions.
- Can distil the essential mathematical information from extended pieces of prose having mathematical content.
- Comment meaningfully on the mathematical information.
- Make appropriate and efficient use of contemporary calculator technology and other permitted resources, and are aware of any limitations to their use.
- Present results to an appropriate degree of accuracy.

Grade C candidates

- Recall or recognise most of the mathematical facts, concepts and techniques that are needed, and usually select appropriate ones to use in a variety of contexts.
- Manipulate mathematical expressions and use graphs, sketches and diagrams, all with a reasonable level of accuracy and skill.
- Use mathematical language with some skill and sometimes proceed logically through extended arguments or proofs.
- When confronted with unstructured problems they sometimes devise and implement an effective and efficient solution strategy.

- Occasionally notice and correct errors in their calculations.
- Recall or recognise most of the standard models that are needed and usually select appropriate ones to represent a variety of situations in the real world.
- Often correctly refer results from calculations using the model to the original situation, they sometimes give sensible interpretations of their results in context of the original realistic situation.
- Sometimes make intelligent comments on the modelling assumptions and possible refinements to the model.
- Comprehend or understand the meaning of most translations into mathematics of common realistic contexts.
- Often correctly refer the results of calculations back to the given context and sometimes make sensible comments or predictions.
- Distil much of the essential mathematical information from extended pieces of prose having mathematical content.
- Give some useful comments on this mathematical information.
- Usually make appropriate and effective use of contemporary calculator technology and other permitted resources, and are sometimes aware of any limitations to their use.
- Usually present results to an appropriate degree of accuracy.

Grade E candidates

- Recall or recognise some of the mathematical facts, concepts and techniques that are needed, and sometimes select appropriate ones to represent to use in some contexts.
- Manipulate mathematical expressions and use graphs, sketches and diagrams, all with some accuracy and skill.
- Sometimes use mathematical language correctly and occasionally proceed logically through extended arguments or proofs.
- Recall or recognise some of the standard models that are needed and sometimes select appropriate ones to represent a variety of situations in the real world.
- Sometimes correctly refer results from calculations using the model to the original situation; they try to interpret their results in the context of the original realistic situation.
- Sometimes comprehend or understand the meaning of translations in mathematics of common realistic contexts.
- Sometimes correctly refer the results of calculations back to the given context and attempt to give comments or predictions.
- Distil some of the essential mathematical information from extended pieces of prose having mathematical content; they attempt to comment on this mathematical information.
- Candidates often make appropriate and efficient use of contemporary calculator technology and other permitted resources.
- Often present results to an appropriate degree of accuracy.

The table below shows how your average mark is translated.

average	*80%*	*70%*	*60%*	*50%*	*40%*
grade	*A*	*B*	*C*	*D*	*E*

Four steps to successful revision

Step 1: Understand

- Study the topic to be learned slowly. Make sure you understand the logic or important concepts.
- Mark up the text if necessary – underline, highlight and make notes.
- Re-read each paragraph slowly.

GO TO STEP 2

Step 2: Summarise

- Now make your own revision note summary:
  What is the main idea, theme or concept to be learned?
  What are the main points? How does the logic develop?
  Ask questions: Why? How? What next?
- Use bullet points, mind maps, patterned notes.
- Link ideas with mnemonics, mind maps, crazy stories.
- Note the title and date of the revision notes
  (e.g. Mathematics: Differentiation, 3rd March).
- Organise your notes carefully and keep them in a file.

This is now in **short-term memory**. You will forget 80% of it if you do not go to Step 3.
GO TO STEP 3, but first take a 10 minute break.

Step 3: Memorise

- Take 25 minute learning 'bites' with 5 minute breaks.
- After each 5 minute break test yourself:
  Cover the original revision note summary
  Write down the main points
  Speak out loud (record on tape)
  Tell someone else
  Repeat many times.

The material is well on its way to **long-term memory**.
You will forget 40% if you do not do step 4. **GO TO STEP 4**

Step 4: Track/Review

- Create a Revision Diary (one A4 page per day).
- Make a revision plan for the topic, e.g. 1 day later, 1 week later, 1 month later.
- Record your revision in your Revision Diary, e.g.
  Mathematics: Differentiation, 3rd March 25 minutes
  Mathematics: Differentiation, 5th March 15 minutes
  Mathematics: Differentiation, 3rd April 15 minutes
  ... and then at monthly intervals.

Chapter 1

Pure 3

The following topics are covered in this chapter:

- Algebra and series
- Coordinate geometry
- Differentiation
- Integration
- Numerical methods
- Vectors

1.1 Algebra and series

After studying this section you should be able to:

- simplify rational expressions
- express rational expressions in partial fractions
- use the binomial expansion $(1+x)^n$ for any rational n

LEARNING SUMMARY

Rational expressions

AQA	P4
EDEXCEL	P3
OCR	P3
WJEC	P3
NICCEA	P3

A **rational expression** is of the form $\frac{f(x)}{g(x)}$, where $f(x)$ and $g(x)$ are polynomials in x.

To simplify a rational expression, factorise both $f(x)$ and $g(x)$ as far as possible, then cancel any factors that appear in both the numerator and denominator.

Examples

(a) $$\frac{2x^2+6x}{2x^2+7x+3} = \frac{2x(x+3)}{(2x+1)(x+3)}$$
$$= \frac{2x}{2x+1}$$

This cannot be cancelled any further.

Factorise $x^2 - 4$ using difference between two squares.

(b) $$\frac{3x^2-12}{2-x} = \frac{3(x^2-4)}{2-x}$$
$$= \frac{3(x-2)(x+2)}{(2-x)}$$
$$= -3(x+2)$$

$x - 2 = -(2 - x)$

When adding and subtracting algebra fractions, always use the lowest common denominator, for **example**,

First factorise the denominators.

The lowest common multiple of the denominators is $(x+3)(x-4)(x-3)$.

$$\frac{x-2}{x^2-x-12} + \frac{4}{x^2-9} = \frac{x-2}{(x+3)(x-4)} + \frac{4}{(x+3)(x-3)}$$
$$= \frac{(x-3)(x-2)+4(x-4)}{(x+3)(x-4)(x-3)}$$
$$= \frac{x^2-5x+6+4x-16}{(x+3)(x-4)(x-3)}$$
$$= \frac{x^2-x-10}{(x+3)(x-4)(x-3)}$$

Partial fractions

AQA	P4
EDEXCEL	P3
OCR	P3
WJEC	P3
NICCEA	P3

It is sometimes possible to decompose a rational function into **partial fractions**. This format is often useful when differentiating, integrating or expanding functions as a series. Look out for the following types of partial fractions.

Key points from AS

- **Identities**
  Revise AS page 27

Denominator contains linear factors only

$$\frac{x+7}{(x-2)(x+1)} \equiv \frac{A}{x-2} + \frac{B}{x+1} = \frac{A(x+1)+B(x-2)}{(x-2)(x+1)}$$

$$\Rightarrow x+7 \equiv A(x+1)+B(x-2)$$

The symbol $\equiv$ indicates that this is an identity. It is true for all values of x.

These substitutions are chosen so that each factor in turn becomes zero.

Let $x = -1$, then $6 = -3B \Rightarrow B = -2$
Let $x = 2$, then $9 = 3A \Rightarrow A = 3$

Examination questions could have up to three linear factors in the denominator.

$$\therefore \frac{x+7}{(x-2)(x+1)} \equiv \frac{3}{x-2} - \frac{2}{x+1}$$

Denominator contains repeated linear factors

Do not use $(x+3)(x+1)(x+1)^2$ as the common denominator as it is not the lowest common multiple.

$$\frac{6x^2+15x+7}{(x+3)(x+1)^2} \equiv \frac{A}{x+3} + \frac{B}{x+1} + \frac{C}{(x+1)^2}$$

$$\equiv \frac{A(x+1)^2+B(x+3)(x+1)+C(x+3)}{(x+3)(x+1)^2}$$

$$\Rightarrow 6x^2+15x+7 \equiv A(x+1)^2+B(x+3)(x+1)+C(x+3)$$

Use a mixture of substitution and equating coefficients to find A, B and C.

Let $x = -1$, then $-2 = 2C \Rightarrow C = -1$
Let $x = -3$, then $16 = 4A \Rightarrow A = 4$
Equate x^2 terms: $6 = A + B$ and since $A = 4$, $B = 2$.

$$\therefore \frac{6x^2+15x+7}{(x+3)(x+1)^2} \equiv \frac{4}{x+3} + \frac{2}{x+1} - \frac{1}{(x+1)^2}$$

Denominator contains a quadratic factor that cannot be factorised

You need to allow for a linear term $(Bx+C)$ above the quadratic factor.

$$\frac{x^2-x+1}{(x-2)(x^2+1)} \equiv \frac{A}{x-2} + \frac{Bx+C}{x^2+1}$$

$$\Rightarrow x^2-x+1 \equiv A(x^2+1)+(Bx+C)(x-2)$$

Using a mixture of substitution and equating coefficients gives $A = \frac{3}{5}$, $B = \frac{2}{5}$, $C = -\frac{1}{5}$.

$$\frac{\frac{3}{5}}{x-2} = \frac{3}{5(x-2)}$$

$$\therefore \frac{x^2-x+1}{(x-2)(x^2+1)} \equiv \frac{3}{5(x-2)} + \frac{2x-1}{5(x^2+1)}$$

Only **proper fractions** (when the numerator is of lower degree than the denominator) can be written in partial fraction form. For improper fractions, divide the denominator into the numerator first. This can be done by long division or by using identities.

The degree of a polynomial is its highest power of x, for example x^3+2x^2-3x+4 has degree 3.

Binomial expansion for rational n

AQA	P5
EDEXCEL	P3
OCR	P3
WJEC	P3
NICCEA	P3

KEY POINT

When n is rational, $(1 + x)^n$ can be written as a series, using the **binomial expansion**

$$(1+x)^n = 1 + nx + \frac{n(n-1)}{2!}x^2 + \frac{n(n-1)(n-2)}{3!}x^3 + \ldots$$

Key points from AS

- **Binomial expansion when n is a positive integer**
  Revise AS page 51

If n is a positive integer, the series terminates at the term in x^n.

For all other values of n, the series is infinite and converges provided that $|x| < 1$.

Example

(a) Expand $\sqrt{1-4x}$ as an ascending series in x, as far as the term in x^3, giving the set of values of x for which the series is valid.

(b) By substituting $x = 0.01$ into the series expansion of $\sqrt{1-4x}$, find $\sqrt{96}$ correct to 4 decimal places.

$1 - 4x = 1 + (-4x)$, so use $(1 + x)^n$ with $n = \frac{1}{2}$ and substituting $(-4x)$ for x.

(a) $\sqrt{1-4x} = (1-4x)^{\frac{1}{2}}$

$$= 1 + \tfrac{1}{2}(-4x) + \frac{(\frac{1}{2})(-\frac{1}{2})}{2!}(-4x)^2 + \frac{(\frac{1}{2})(-\frac{1}{2})(-\frac{3}{2})}{3!}(-4x)^3 + \ldots$$

$$= 1 - 2x - 2x^2 - 4x^3 + \ldots$$

You should also state the set of values of x for which the expansion is valid, even when it is not specifically requested in the question.

The series is valid provided that $|4x| < 1$, i.e. $|x| < \frac{1}{4}$.

(b) Substituting $x = 0.01$ into $\sqrt{1-4x}$ gives $\sqrt{0.96} = \sqrt{\dfrac{96}{100}} = \dfrac{\sqrt{96}}{10}$.

So $\sqrt{96} = 10(1 - 2(0.01) - 2(0.01)^2 - 4(0.01)^3 + \ldots)$
$= 10(1 - 0.02 - 0.0002 - 0.000004 +)$
$= 9.7980$ (4 d.p.)

The function to be expanded must be in the form $(\mathbf{1} + \ldots)^n$ or $(\mathbf{1} - \ldots)^n$.
You may have to do some careful algebraic manipulation to get it in this form.

Example

Give the first four terms in the series expansion of $\dfrac{1}{(2+x)}$.

$$\frac{1}{(2+x)} = \frac{1}{2\left(1+\frac{x}{2}\right)}$$

$$= \frac{1}{2}\left(1+\frac{x}{2}\right)^{-1}$$

Alternatively, $(2+x)^{-1} = 2^{-1}\left(1+\dfrac{x}{2}\right)^{-1}$.

$$= \frac{1}{2}\left(1 + (-1)\left(\frac{x}{2}\right) + \frac{(-1)(-2)}{2!}\left(\frac{x}{2}\right)^2 + \frac{(-1)(-2)(-3)}{3!}\left(\frac{x}{2}\right)^3 + \ldots\right.$$

$$= \frac{1}{2}\left(1 - \frac{x}{2} + \frac{x^2}{4} - \frac{x^3}{8} + \ldots\right)$$

$$= \frac{1}{2} - \frac{x}{4} + \frac{x^2}{8} - \frac{x^3}{16} + \ldots$$

Restriction on x: $\left|\dfrac{x}{2}\right| < 1$, i.e. $|x| < 2$.

Progress check

1 Simplify:

(a) $\dfrac{2x^2+3x}{2x^2+x-3}$ (b) $\dfrac{9x-x^3}{3x^2+x^3}$

2 Express as a single fraction in its simplest form:

(a) $\dfrac{3(x-4)}{(x+2)(x-1)}+\dfrac{2(x+1)}{x-1}$

(b) $\dfrac{x^2-3x+2}{x^2+3x-4}-\dfrac{1}{(x+4)^2}$

3 $\dfrac{4x^2+x+1}{(x+2)(x-3)(x+1)} \equiv \dfrac{A}{x+2}+\dfrac{B}{x-3}+\dfrac{C}{x+1}$

Find the values of A, B and C.

4 Express in partial fractions:

(a) $\dfrac{3x^2+4x+5}{(x-1)^2(x+2)}$ (b) $\dfrac{3x+1}{(x^2+1)(x-3)}$

5 Write as a series in ascending powers of x as far as the term in x^3, stating the values of x for which the expansion is valid:

(a) $\dfrac{1}{1+x}$ (b) $(1-3x)^{-2}$ (c) $\sqrt{4+x}$

1 (a) $\dfrac{x-1}{x}$ (b) $\dfrac{3-x}{x}$

2 (a) $\dfrac{2x^2+9x-8}{(x+2)(x-1)}$ (b) $\dfrac{x^2+2x-9}{(x+4)^2}$

3 $A = 3$, $B = 2$, $C = -1$

4 (a) $\dfrac{2}{x-1}+\dfrac{4}{(x-1)^2}+\dfrac{1}{x+2}$ (b) $\dfrac{1}{x-3}-\dfrac{x}{x^2+1}$

5 (a) $1-x+x^2-x^3+\dots$ for $|x|<1$

(b) $1+6x+27x^2+108x^3+\dots$ for $|x|<\frac{1}{3}$

(c) $2+\frac{1}{4}x-\frac{1}{64}x^2+\frac{1}{512}x^3+\dots$ for $|x|<4$

1.2 Coordinate geometry

After studying this section you should be able to:

- *sketch curves in Cartesian and parametric form*
- *convert the equation of a curve between Cartesian and parametric form*
- *find the centre and radius of a circle from its equation*
- *solve problems involving lines and circles*
- *find the angle between two lines*

LEARNING SUMMARY

Cartesian and parametric form

AQA	P5
EDEXCEL	P3
OCR	P3
WJEC	P3
NICCEA	P3

Key points from AS

- **Curve sketching**
  Revise AS page 39

Cartesian form

A curve in Cartesian form is defined in terms of x and y only, **for example**
$y = x^2 + 3x$ in which y is given **explicitly** in terms of x.
$x^2 + 2xy + y^2 = 9$ in which the equation is given **implicitly**.

To sketch a function given in Cartesian form:

If all the powers of x are even, the graph is symmetrical in the y-axis.

Look for values that make the denominator zero.

1 Find where the curve **crosses the axes** by putting $x = 0$ and $y = 0$.
2 Check for **symmetry**.
3 Check for **stationary points**.
4 Check what happens for **large values of x and y**.
5 Check for **discontinuities**.

Find when $\frac{dy}{dx} = 0$ and then investigate.

Parametric form

The parameter θ is often used when trigonometric functions are involved.

A curve is defined in **parametric form** by expressing x and y in terms of a third variable, for example $x = f(t)$, $y = g(t)$. In this case, t is the parameter.

The curve is plotted by finding the values of x and y for various values of the parameter.

Example
Plot the curve $x = 2t^2$, $y = 4t$ for $-3 \leqslant t \leqslant 3$ and find the Cartesian equation of the curve.

The curve is a parabola.

t	−3	−2	−1	0	1	2	3
x	18	8	2	0	2	8	18
y	−12	−8	−4	0	4	8	12

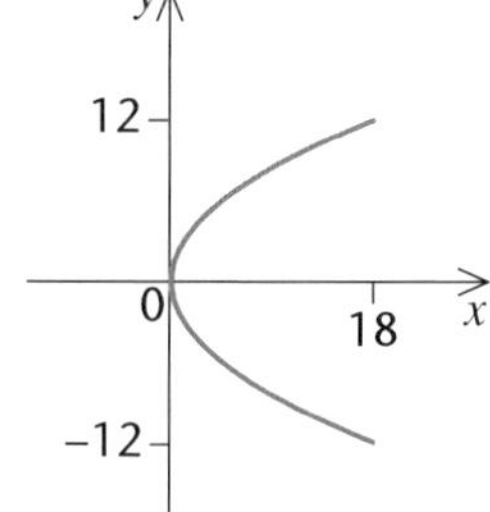

Eliminate t to obtain the Cartesian equation.

$y = 4t \Rightarrow t = \frac{y}{4}$.

Key points from AS

- **Trigonometric identities**
  Revise AS page 27

Substituting into $x = 2t^2$ gives $x = 2\left(\frac{y}{4}\right)^2 = \frac{y^2}{8}$ so $y^2 = 8x$.

When the parameter is θ, it may be necessary to use a trigonometric identity to find the Cartesian equation.

For example, when $x = a\cos\theta$, $y = b\sin\theta$, use

This curve is an ellipse.

$\cos^2\theta + \sin^2\theta = 1$. This gives $\frac{x^2}{a^2} + \frac{y^2}{b^2} = 1$.

Circles

AQA	P4
EDEXCEL	P3
OCR	P3
WJEC	P2
NICCEA	P3

KEY POINT

The equation of a **circle**, centre (a, b), radius r is

$$(x-a)^2+(y-b)^2=r^2.$$

For **example**

In the equation of a circle, the coefficients of x^2 and y^2 are the same. The equation may contain an x-term, a y-term and a constant, but no other terms (such as xy).

The equation of a circle with centre (3, –5) and radius 4 is $(x-3)^2+(y+5)^2=16$.

Expanding this equation gives $x^2-6x+9+y^2+10y+25=16$,
i.e. $x^2+y^2-6x+10y+18=0$.

An equation in this form can be rearranged to find the centre and radius. In general

To find the centre and radius, complete the square for the x-terms and for the y-terms:

$$x^2+y^2+2gx+2fy+c=0$$

i.e. $$x^2+2gx+y^2+2fy+c=0$$

$$\Rightarrow\quad (x+g)^2-g^2+(y+f)^2-f^2+c=0,$$

$$(x+g)^2+(y+f)^2=g^2+f^2-c$$

Compare with $(x-a)^2+(y-b)^2=r^2$

The centre is $(-g,-f)$ and the radius is $\sqrt{g^2+f^2-c}$.

A circle, radius r, centre (0, 0), has Cartesian form $x^2+y^2=r^2$ and parametric form $x=r\cos\theta$, $y=r\sin\theta$, for $0\leqslant\theta\leqslant 2\pi$.

Useful **geometrical facts:**

You should learn these.

- The angle in a semicircle is 90°.
- The angle between a tangent and radius is 90°.
- Tangents to a circle from an external point are equal in length.
- The shortest distance from the centre to a chord is the perpendicular distance.
- The perpendicular from the centre to a chord bisects the chord.

Key points from AS

- **Coordinate geometry**
  Revise AS page 31

Example

(a) The circle $x^2+y^2+2x-4y-20=0$ has centre C and radius r.
Show that $r=5$ and find the coordinates of C.

(b) Find the equation of the tangent at P (2, 6)

Complete the square.

(a) $$x^2+2x\quad+y^2-4y\quad-20=0$$

$$(x+1)^2-1+(y-2)^2-4-20=0$$

Compare with $(x-a)^2+(y-b)^2=r^2$

$$\Rightarrow\qquad (x+1)^2+(y-2)^2=25$$

Centre C is (–1, 2) and radius is 5.

For perpendicular lines, product of gradients = –1.

(b) gradient $CP=\dfrac{6-2}{2-(-1)}=\dfrac{4}{3}$.

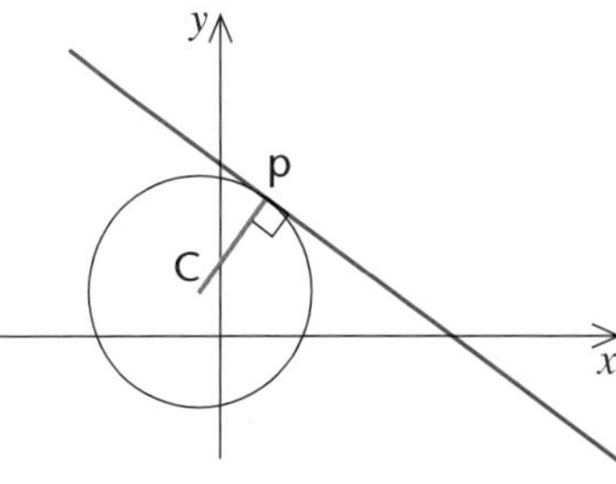

An alternative method using calculus is shown on p. 29.

Since the tangent is perpendicular to the radius, gradient of tangent at P is $-\frac{3}{4}$.

Use $y-y_1=m(x-x_1)$

Equation of tangent at P:

$$y-6=-\tfrac{3}{4}(x-2)$$

$$\Rightarrow 4y-24=-3x+6$$

$$\Rightarrow 4y=-3x+30$$

Angle between two lines

AQA P4

KEY POINT

The angle θ between two straight lines with gradients m_1 and m_2 is given by

$$\tan\theta = \frac{m_1 - m_2}{1 + m_1 m_2}.$$

The two lines are parallel if $m_1 = m_2$.
The two lines are perpendicular if $m_1 m_2 = -1$

Example

Find the angle between the two lines $y = 4x - 2$ and $2y - 3x = 1$.

$2y - 3x = 1$
$\Rightarrow 2y = 3x + 1$
$\Rightarrow y = 1.5x + 0.5$

Gradients: $m_1 = 4$, $m_2 = 1.5$

$$\tan\theta = \frac{m_1 - m_2}{1 + m_1 m_2} = \frac{4 - 1.5}{1 + 4 \times 1.5} = 0.3571\ \ldots$$

$\Rightarrow \quad \theta = 19.7°$ (1 d.p.)

Progress check

1 Sketch these curves:

(a) $y = x^3 + 3x^2$ (b) $y = \dfrac{1}{2x - 4}$ (c) $y = \dfrac{x^2 + 1}{x}$

2 A curve is defined parametrically by $x = 2t$, $y = \dfrac{2}{t}$.

(a) Sketch the curve.
(b) Give the Cartesian equation of the curve.

3 (a) Find in Cartesian form the equation of the circle C with centre (2, –3) and radius 6.
(b) Determine, by calculation whether (6, 1) lies inside or outside the circle.

4 Find the centre and radius of the circle with equation:
(a) $x^2 + y^2 + 8x + 2y - 5 = 0$
(b) $2x^2 + 2y^2 - 6x + 3 = 0$.

(b) Divide throughout by 2 first.

5 Find the acute angle between the lines $2y + 6x = 9$ and $y = 5x - 3$.

1 (a)

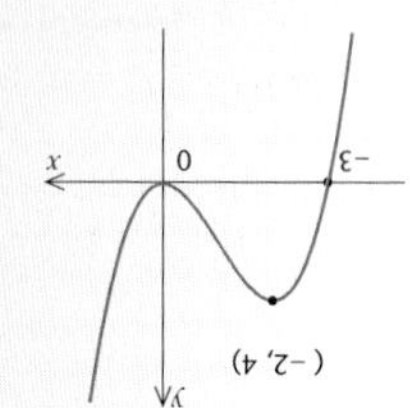

(b) (c)

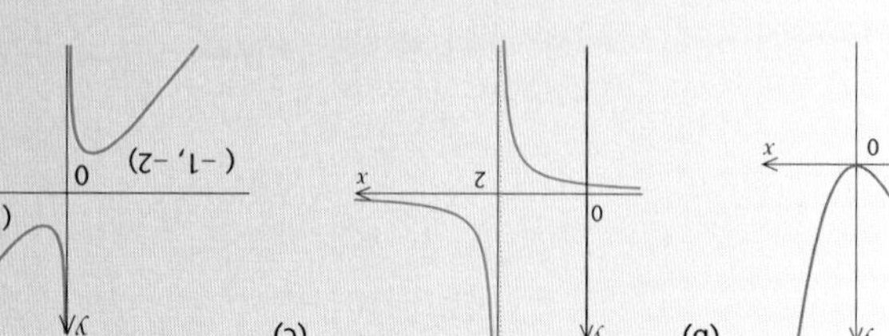

2 (a)

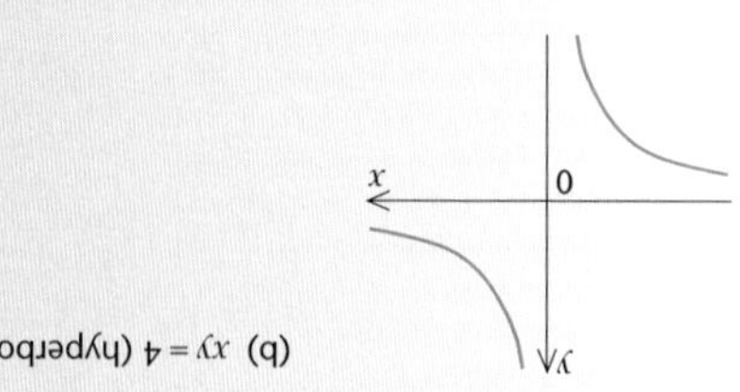

(b) $xy = 4$ (hyperbola)

3 (a) $(x-2)^2 + (y+3)^2 = 36 \Rightarrow x^2 + y^2 - 4x + 6y - 23 = 0$ (b) Inside

4 (a) $(-4, -1)$, $\sqrt{22}$ (b) $(1.5, 0)$, $\dfrac{\sqrt{3}}{2}$

5 29.7°

1.3 Differentiation

After studying this section you should be able to:

- differentiate trigonometric functions
- differentiate logarithmic functions
- differentiate using the product and quotient rules
- find the first and second derivatives of functions defined parametrically and implicitly
- find approximate changes
- investigate points of inflexion

LEARNING SUMMARY

Trigonometric functions

AQA	P4
EDEXCEL	P3
OCR	P3
WJEC	P2
NICCEA	P3

You should learn the derivatives of the three main **trigonometric functions**:

$$\frac{d}{dx}(\sin ax) = a\cos ax$$
$$\frac{d}{dx}(\cos ax) = -a\sin ax$$
$$\frac{d}{dx}(\tan ax) = a\sec^2 ax$$

Example

When differentiating, the angle must be in radians.

Find the exact value of the gradient of $y = \cos 2x$ when $x = \frac{\pi}{6}$.

$$y = \cos 2x \Rightarrow \frac{dy}{dx} = -2\sin 2x$$

When $x = \frac{\pi}{6}$, $\frac{dy}{dx} = -2\sin\frac{\pi}{3} = -2 \times \frac{\sqrt{3}}{2} = -\sqrt{3}$

Key points from AS

- **Trigonometry** *Revise AS pages 37*
- **Diffentiation** *Revise AS pages 38–40*
- **Chain rule** *Revise AS pages 60–61*

The **chain rule** can often be used to differentiate expressions involving trigonometric functions.

Example

Differentiate: (a) $y = 5\sin^4 x$ (b) $y = e^{\tan 4x}$

$\sin^4 x = (\sin x)^4$

(a) Let $u = \sin x$ so that $y = 5u^4$

Then $\frac{du}{dx} = \cos x$ and $\frac{dy}{du} = 20u^3 = 20(\sin x)^3 = 20\sin^3 x$

With practice you should be able to do the working mentally.

By the chain rule:

$$\frac{dy}{dx} = \frac{dy}{du} \times \frac{du}{dx} = 20\sin^3 x \times (\cos x)$$
$$= 20\cos x \sin^3 x.$$

Although not essential, it is usual to write an expression with the least complicated factors first.

(b) Let $u = \tan 4x$ so that $y = e^u$

$\frac{du}{dx} = 4\sec^2 4x$ and $\frac{dy}{du} = e^u = e^{\tan 4x}$.

$$\therefore \frac{dy}{dx} = e^{\tan 4x} \times 4\sec^2 4x$$
$$= 4\sec^2 4x\, e^{\tan 4x}.$$

Key points from AS

- **Exponential function** *Revise AS page 60*
- **Trigonometry** *Revise AS page 56*

In general, by the chain rule: $\frac{d}{dx}(e^{f(x)}) = f'(x)e^{f(x)}$

$\sec x = \frac{1}{\cos x} = (\cos x)^{-1}$

Do not confuse $(\cos x)^{-1}$ with the inverse function arc cos x, often written $\cos^{-1} x$.

The chain rule can also be used to differentiate sec x, cosec x and cot x, for **example** if $y = \sec x = (\cos x)^{-1}$

then $\frac{dy}{dx} = -(\cos x)^{-2}(-\sin x)$

$$= \frac{\sin x}{\cos x \cos x} = \frac{1}{\cos x} \times \frac{\sin x}{\cos x} = \sec x \tan x.$$

In general:

These should be in the examination formulae booklet, but it is a good idea to be able to derive them yourself.

$$\frac{d}{dx}(\sec x) = \sec x \tan x$$

$$\frac{d}{dx}(\text{cosec } x) = -\text{cosec } x \cot x$$

$$\frac{d}{dx}(\cot x) = -\text{cosec}^2 x$$

Examples

Use the chain rule
(a) Let $u = 5x$
(b) Let $u = x^2$

(a) $\frac{d}{dx}(4 \cot 5x) = -20 \text{ cosec}^2 5x$

(b) $\frac{d}{dx}(\text{cosec}(x^2)) = -2x \text{ cosec}(x^2)\cot(x^2)$

Logarithmic functions

AQA	P4
EDEXCEL	P3
OCR	P2
WJEC	P2
NICCEA	P2

Example

Differentiate $y = \ln(x^2 + 3x)$

Let $u = x^2 + 3x$ so that $y = \ln u$

$$\frac{du}{dx} = 2x + 3 \quad \text{and} \quad \frac{dy}{du} = \frac{1}{u} = \frac{1}{x^2 + 3x}$$

Using the chain rule

$$\frac{dy}{dx} = \frac{dy}{du} \times \frac{du}{dx} = \frac{1}{x^2 + 3x} \times (2x + 3)$$

$$= \frac{2x + 3}{x^2 + 3x}$$

Notice that $2x + 3$ is the derivative of $x^2 + 3x$.

In general, it can be shown using the chain rule

This can be used directly without showing the chain rule working.

KEY POINT

$$\frac{d}{dx}\ln(f(x)) = \frac{f'(x)}{f(x)}$$

Example

$f(x) = \sin x, f'(x) = \cos x$

(a) $\frac{d}{dx}\ln(\sin x) = \frac{\cos x}{\sin x} = \cot x$

Always simplify a log expression first, if possible.

Use $\ln a^n = n \ln a$

(b) $\frac{d}{dx}\ln(\sqrt{4 + x^3}) = \frac{d}{dx}(\tfrac{1}{2}\ln(4 + x^3))$

$$= \frac{1}{2}\frac{d}{dx}\ln(4 + x^3)$$

$$= \frac{1}{2}\left(\frac{3x^2}{4 + x^3}\right) = \frac{3x^2}{2(4 + x^3)}$$

Key points from AS

- **Logarithms**
  Revise AS page 55

The product rule

AQA P3
EDEXCEL P3
OCR P3
WJEC P2
NICCEA P4

Expressions written as the product of two functions, $f(x) \times g(x)$, can be differentiated using the **product rule**.

KEY POINT

If $y = uv$, where $u = f(x)$ and $v = g(x)$, then $\dfrac{dy}{dx} = u\dfrac{dv}{dx} + v\dfrac{du}{dx}$.

Example

Differentiate with respect to x: (a) $y = x^2 \sin 4x$ (b) $y = x^3 e^{-x}$

With practice, the side working can be done mentally.

(a) Let $y = uv$, where $u = x^2$ and $v = \sin 4x$

$$\frac{dy}{dx} = u\frac{dv}{dx} + v\frac{du}{dx}$$

$$= x^2 \times 4\cos 4x + \sin 4x \times 2x$$

$$= 2x(2x\cos 4x + \sin 4x)$$

Side working

$u = x^2 \Rightarrow \dfrac{du}{dx} = 2x$

$v = \sin 4x \Rightarrow \dfrac{dv}{dx} = 4\cos 4x$

Tidy the answer by taking out factors if possible.

(b) Let $y = uv$, where $u = x^3$ and $v = e^{-x}$

$$\frac{dy}{dx} = x^3 \times (-e^{-x}) + e^{-x} \times 3x^2$$

$$= x^2 e^{-x}(-x + 3)$$

Side working

$u = x^3 \Rightarrow \dfrac{du}{dx} = 3x^2$

$v = e^{-x} \Rightarrow \dfrac{dv}{dx} = -e^{-x}$

The quotient rule

AQA P3
EDEXCEL P3
OCR P3
WJEC P2
NICCEA P4

Expressions written in the form $\dfrac{f(x)}{g(x)}$ can be differentiated using the **quotient rule**.

KEY POINT

If $y = \dfrac{u}{v}$ where $u = f(x)$ and $v = g(x)$, then $\dfrac{dy}{dx} = \dfrac{v\dfrac{du}{dx} - u\dfrac{dv}{dx}}{v^2}$

Before using the quotient rule, check whether the expression can be simplified first, such as in $y = \dfrac{4x^2 + 3x + 2}{x}$

Example

Find $\dfrac{dy}{dx}$ when $y = \dfrac{e^{2x}}{x^3}$.

$$\frac{dy}{dx} = \frac{v\dfrac{du}{dx} - u\dfrac{dv}{dx}}{v^2}$$

$$= \frac{x^3(2e^{2x}) - e^{2x}(3x^2)}{(x^3)^2}$$

Take out factors.

$$= \frac{x^2 e^{2x}(2x - 3)}{x^6}$$

Simplify.

$$= \frac{e^{2x}(2x - 3)}{x^4}$$

Side working:

$u = e^{2x} \Rightarrow \dfrac{du}{dx} = 2e^{2x}$

$v = x^3 \Rightarrow \dfrac{dv}{dx} = 3x^2$

Example

Find the coordinates of the stationary point on the curve $y = \frac{\ln x}{x}$.

Side working:

$$u = \ln x \Rightarrow \frac{du}{dx} = \frac{1}{x}$$

$$v = x \Rightarrow \frac{dv}{dx} = 1$$

$$\frac{dy}{dx} = \frac{x \times \frac{1}{x} - \ln x \times 1}{x^2} = \frac{1 - \ln x}{x^2}$$

Key points from AS

- **Stationary points** *Revise AS page 39*

$\frac{dy}{dx} = 0$ when $\ln x = 1$, i.e. when $x = e$

$\ln e = 1$

When $x = e$, $y = e^{-1}$, so the stationary point is at (e, e^{-1}).

Alternative formats of product and quotient rules when $u = f(x)$ and $v = g(x)$:

These formats are sometimes quoted. Learn and use the ones you find easiest to remember.

Product rule: If $y = f(x) \times g(x)$, then $\frac{dy}{dx} = f'(x)g(x) + f(x)g'(x)$

Quotient rule: If $y = \frac{f(x)}{g(x)}$, then $\frac{dy}{dx} = \frac{f'(x)g(x) - f(x)g'(x)}{[g(x)]^2}$

Parametric functions

First derivative

AQA	P5
EDEXCEL	P3
OCR	P3
WJEC	P3
NICCEA	P3

Second derivative

AQA	P5
NICCEA	P3

KEY POINT

If x and y are each expressed in terms of a **parameter** t, then $\frac{dy}{dx} = \frac{dy}{dt} \times \frac{dt}{dx}$ Remember that $\frac{dt}{dx} = \frac{1}{dx/dt}$

Example

A curve is defined parametrically by $x = 3t^2$, $y = 4t$.

(a) Find $\frac{dy}{dx}$ when $y = 8$ (b) Find the value of $\frac{d^2y}{dx^2}$ when $t = -1$.

(a) $x = 3t^2 \Rightarrow \frac{dx}{dt} = 6t \Rightarrow \frac{dt}{dx} = \frac{1}{6t}$

$y = 4t \Rightarrow \frac{dy}{dt} = 4$

$\therefore \frac{dy}{dx} = \frac{dy}{dt} \times \frac{dt}{dx} = 4 \times \frac{1}{6t} = \frac{2}{3t}$

Find the value of t when $y = 8$.

When $y = 8$, $4t = 8$, i.e. $t = 2$; when $t = 2$, $\frac{dy}{dx} = \frac{2}{3t} = \frac{2}{6} = \frac{1}{3}$.

If $\frac{dy}{dx} = h(t)$, then $\frac{d^2y}{dx^2} = \frac{d}{dt}(h(t)) \times \frac{dt}{dx}$

(b) $\frac{d^2y}{dx^2} = \frac{d}{dx}\left(\frac{2}{3t}\right)$

$\frac{d}{dt}\left(\frac{2}{3t}\right) \times \frac{dt}{dx} =$

$= -\frac{2}{3}t^{-2} \times \frac{1}{6t} = -\frac{1}{9t^3}$

When $t = -1$, $\frac{d^2y}{dx^2} = \frac{1}{9}$

Implicit functions

First derivative

AQA	P5
EDEXCEL	P3
OCR	P3
WJEC	P3
NICCEA	P3

Second derivative

AQA	P5
NICCEA	P3

To find $\frac{dy}{dx}$ when an equation in x and y is given **implicitly**, differentiate each term with respect to x, remembering that $\frac{d}{dx}(f(y)) = \frac{d}{dy}f(y) \times \frac{dy}{dx}$.

Example

The point $P(2, 6)$ lies on the circle $x^2 + y^2 + 2x - 4y - 20 = 0$.

Compare this with the method on p. 23.

(a) Show that $\frac{dy}{dx} = \frac{x+1}{y-2}$

(b) Calculate the gradient of the tangent at P.

(c) Calculate the value of $\frac{d^2y}{dx^2}$ at P.

Differentiate term by term with respect to x: $\frac{d}{dx}(y) = \frac{dy}{dx}$

(a) $2x + 2y\frac{dy}{dx} + 2 - 4\frac{dy}{dx} + 0 = 0$

Divide throughout by 2 to simplify the equation.

$$x + y\frac{dy}{dx} + 1 - 2\frac{dy}{dx} = 0 \qquad (1)$$

Get the terms in $\frac{dy}{dx}$ on one side of the equation and all the other terms on the other side.

$$x + 1 = 2\frac{dy}{dx} - y\frac{dy}{dx}$$

Take out $\frac{dy}{dx}$ as a factor.

$$\Rightarrow x + 1 = \frac{dy}{dx}(2 - y)$$

$$\therefore \frac{dy}{dx} = \frac{x+1}{2-y}$$

The value of $\frac{dy}{dx}$ when $x = 2$ gives the gradient of the tangent at (2, 6)

(b) At P, $x = 2$ and $y = 6$,

so $\frac{dy}{dx} = \frac{3}{-4} = -\frac{3}{4}$.

The gradient of the tangent at P is $-\frac{3}{4}$.

To find $\frac{d^2y}{dx^2}$, differentiate (1) with respect to x.

(c)
$$x + y\frac{dy}{dx} + 1 - 2\frac{dy}{dx} = 0 \qquad \text{from (1) above}$$

Use the product rule to differentiate $y\frac{dy}{dx}$.

$$1 + \left(y\frac{d^2y}{dx^2} + \frac{dy}{dx} \times \frac{dy}{dx}\right) - 2\frac{d^2y}{dx^2} = 0$$

$$1 + (y-2)\frac{d^2y}{dx^2} + \left(\frac{dy}{dx}\right)^2 = 0 \qquad (2)$$

At P, $y = 6$ and $\frac{dy}{dx} = -\frac{3}{4}$ from part (b)

Substituting in (2)

$$1 + 4\frac{d^2y}{dx^2} + \left(-\frac{3}{4}\right)^2 = 0$$

$$\Rightarrow \frac{d^2y}{dx^2} = -\frac{25}{64}$$

Approximate changes

AQA P4

For two variables x and y related by the equation $y = f(x)$:

$\lim_{\delta x \to 0} \frac{\delta y}{\delta x} = \frac{dy}{dx}$

KEY POINT

If δx is small, $\frac{\delta y}{\delta x} \approx \frac{dy}{dx} \Rightarrow \delta y \approx \frac{dy}{dx}\delta x$.

Example

A 3% error is made when measuring the radius of a sphere. Using calculus, find the approximate percentage error that results when calculating the volume.

This calculus method gives an approximation. The smaller the value of δr, the better the approximation.

The error in r is 3%, so $\delta r = 0.03r$.

Volume of sphere $V = \frac{4}{3}\pi r^3 \Rightarrow \frac{dV}{dr} = 4\pi r^2$.

If r is small, $\frac{\delta V}{\delta r} \approx \frac{dV}{dr}$

$\Rightarrow \qquad \delta V \approx \frac{dV}{dr}\delta r$

so $\qquad \delta V \approx 4\pi r^2 \delta r = 4\pi r^2 \times 0.03r = 0.12\pi r^3$

Percentage error when calculating $= \frac{\delta V}{V} \times 100 \approx \frac{0.12\pi r^3}{\frac{4}{3}\pi r^3} \times 100 = 9\%$.

Points of inflexion

AQA P5

If, at a particular point on the curve $y = f(x)$, $\frac{d^2y}{dx^2} = 0$ and $\frac{d^3y}{dx^3} \neq 0$, then the point is a **point of inflexion**. If $\frac{d^2y}{dx^2} = 0$ and $\frac{d^3y}{dx^3} = 0$, further investigation is required.

Different types of points of inflexion are illustrated in the following diagrams:

If the gradient at the point of inflexion is zero, then it is also a stationary point.

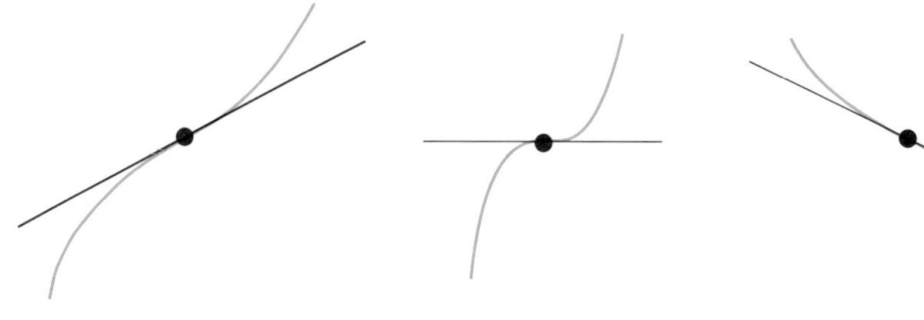

Example

Find the x coordinates of any points of inflexion on the curve $y = \frac{1}{4}x^4 - x^3 + 2x + 1$.

$\frac{dy}{dx} = x^3 - 3x^2 + 2$

$\frac{d^2y}{dx^2} = 3x^2 - 6x = 3x(x-2)$

$\frac{d^3y}{dx^3} = 6x - 6$

$\frac{d^2y}{dx^2} = 0$ when $3x(x-2) = 0$, i.e. when $x = 0$ and $x = 2$.

When $x = 0$, $\frac{d^3y}{dx^3} \neq 0$ when $x = 2$, $\frac{d^3y}{dx^3} \neq 0$,

There are points of inflexion when $x = 0$ and $x = 2$.

Progress check

1 Differentiate with respect to x:

(a) $y = 2 \sin 6x$ (b) $y = 4 \cos^3 x$ (c) $y = 2 \tan 3x$

(d) $y = e^{\sin x}$ (e) $y = \sin(x^3)$ (f) $y = \dfrac{1}{\tan 3x}$

(f) Re-write in terms of cot $3x$

Use a trigonometric identity.

2 If $y = \sin^2 x$, show that $\dfrac{dy}{dx} = \sin 2x$.

3 Differentiate with respect to x:

(a) $\ln(x^2 + 3)$ (b) $\ln((2x - 4)^5)$ (c) $\ln(\cos 2x)$ (d) $\ln(6x - 1)$

4 Find $\dfrac{dy}{dx}$, simplifying your answers.

(a) $y = x^2 \cos 3x$ (b) $y = e^{2x} \sin x$ (c) $y = x \ln x$ (d) $y = (x + 1)(2x - 3)^4$

5 Find the equation of the tangent to the curve $y = x\sqrt{1 + 2x}$ at the point (4, 12).

6 Use the quotient rule to find $\dfrac{dy}{dx}$, simplifying your answers where necessary.

(a) $y = \dfrac{\sin 2x}{x}$ (b) $y = \dfrac{2x - 1}{3x + 4}$ (c) $y = \dfrac{\ln(x + 2)}{x + 2}$

7 A curve is defined parametrically by $x = t^2$, $y = t^3$.

(a) Find $\dfrac{dy}{dx}$ and $\dfrac{d^2y}{dx^2}$ when $t = 2$.

(b) Find the Cartesian equation of the curve.

8 A curve has equation $x^2 + xy^2 - 5 = 0$.

(a) Find the value of $\dfrac{dy}{dx}$ at (1, 2).

(b) Find the equation of the normal when $x = 1$, in the form $ax + by + c = 0$.

1 (a) $12 \cos 6x$ (b) $-12 \sin x \cos^2 x$ (c) $6 \sec^2 3x$ (d) $\cos x e^{\sin x}$
(e) $3x^2 \cos(x^3)$ (f) $-3 \operatorname{cosec}^2 3x$

3 (a) $\dfrac{2x}{x^2 + 3}$ (b) $\dfrac{5}{x - 2}$ (c) $-2 \tan 2x$ (d) $\dfrac{6}{6x - 1}$

4 (a) $x(2 \cos 3x - 3x \sin 3x)$ (b) $e^{2x}(\cos x + 2 \sin x)$
(c) $1 + \ln x$ (d) $5(2x + 1)(2x - 3)^3$

5 $3y = 13x - 16$

6 (a) $\dfrac{2x \cos 2x - \sin 2x}{x^2}$ (b) $\dfrac{11}{(3x + 4)^2}$ (c) $\dfrac{1 - \ln(x + 2)}{(x + 2)^2}$

7 (a) $3, \frac{3}{8}$ (b) $x^3 = y^2$

8 (a) -1.5 (b) $2x - 3y + 4 = 0$

1.4 Integration

After studying this section you should be able to:

LEARNING SUMMARY

- integrate trigonometrical functions
- recognise an integral that leads to a logarithmic function
- integrate using partial fractions
- integrate using a substitution
- use integration by parts to integrate a product
- find the area under a curve defined parametrically
- solve first order differential equations when the variables can be separated
- understand exponential growth and decay

Trigonometric functions

AQA	P4
EDEXCEL	P3
OCR	P3
WJEC	P2
NICCEA	P3

The following integrals are obtained by applying the reverse process of differentiating the three main trigonometric functions (page 25).

You should learn these and practise using them.

$$\int \cos(ax)\mathrm{d}x = \frac{1}{a}\sin(ax) + c$$

$$\int \sin(ax)\mathrm{d}x = -\frac{1}{a}\cos(ax) + c$$

$$\int \sec^2(ax)\mathrm{d}x = \frac{1}{a}\tan(ax) + c$$

Example

It is a good idea to check your answer by differentiating.

$$\int (4\cos 2x - \sin 2x)\mathrm{d}x = 2\sin 2x + \tfrac{1}{2}\cos 2x + c$$

Example

When integrating, the angle must be in radians.

$$\int_{\frac{\pi}{3}}^{\frac{\pi}{2}} \sec^2(\tfrac{1}{2}x)\mathrm{d}x = \left[\frac{1}{\frac{1}{2}}\tan(\tfrac{1}{2}x)\right]_{\frac{\pi}{3}}^{\frac{\pi}{2}}$$

$$= 2\left[\tan(\tfrac{1}{2}x)\right]_{\frac{\pi}{3}}^{\frac{\pi}{2}}$$

$$= 2\left(\tan\frac{\pi}{4} - \tan\frac{\pi}{6}\right)$$

$$= 2\left(1 - \frac{1}{\sqrt{3}}\right)$$

Learn the exact values of the trigonometric ratios of $\frac{\pi}{6}, \frac{\pi}{4}, \frac{\pi}{3}$ (30°, 45°, 60°)

These results are very useful and can be recognised easily:

$$\frac{\mathrm{d}}{\mathrm{d}x}(\sin^{n+1}x) = (n+1)\sin^n x\cos x \quad \Rightarrow \quad \int \sin^n x\cos x\mathrm{d}x = \frac{1}{n+1}\sin^{n+1}x + c$$

$$\frac{\mathrm{d}}{\mathrm{d}x}(\cos^{n+1}x) = -(n+1)\cos^n x\sin x \quad \Rightarrow \quad \int \cos^n x\sin x\mathrm{d}x = -\frac{1}{n+1}\cos^{n+1}x + c$$

Key points from AS

- **Trigonometry**
  Revise AS page 35

Examples

$$\int \sin^5 x\cos x\mathrm{d}x = \tfrac{1}{6}\sin^6 x + c$$

$$\int \sin x\cos^3 x\mathrm{d}x = -\tfrac{1}{4}\cos^4 x + c$$

Odd powers of sin x and cos x

Example

Evaluate $\int_0^{\frac{\pi}{6}} \cos^3 x\mathrm{d}x$

Use $\cos^2 x = 1 - \sin^2 x$

Recognise the integral of $\cos x \sin^2 x$

$$\begin{aligned}\int_0^{\frac{\pi}{6}} \cos^3 x\mathrm{d}x &= \int_0^{\frac{\pi}{6}} \cos x(\cos^2 x)\mathrm{d}x \\ &= \int_0^{\frac{\pi}{6}} \cos x(1 - \sin^2 x)\mathrm{d}x \\ &= \int_0^{\frac{\pi}{6}} (\cos x - \cos x \sin^2 x)\mathrm{d}x \\ &= \left[\sin x - \tfrac{1}{3}\sin^3 x\right]_0^{\frac{\pi}{6}} \\ &= \tfrac{1}{2} - \tfrac{1}{24} = \tfrac{11}{24}\end{aligned}$$

Key points from AS

- **Double angle formulae**
  Revise AS page 58

Even powers of cos x and sin x

To integrate even powers of sin x and cos x, use the double angle identities for cos $2x$.

$$\cos 2x = 2\cos^2 x - 1 \quad \Rightarrow \quad \cos^2 x = \tfrac{1}{2}(1 + \cos 2x)$$
$$\cos 2x = 1 - 2\sin^2 x \quad \Rightarrow \quad \sin^2 x = \tfrac{1}{2}(1 - \cos 2x)$$

Example

Try $\int \sin^2 x\mathrm{d}x$

$$\begin{aligned}\int \cos^2 x\mathrm{d}x &= \frac{1}{2}\int (1 + \cos 2x)\mathrm{d}x \\ &= \tfrac{1}{2}(x + \tfrac{1}{2}\sin 2x) + c\end{aligned}$$

To integrate $\cos^4 x$, write it as $(\cos^2 x)^2 = (\tfrac{1}{2}(1 + \cos 2x))^2$.

The following integrals can be deduced by differentiating sec x, cosec x and cot x:

It is likely that these will be given in the examination formulae booklet but it is a good idea to practise using them.

$$\int \sec x \tan x\mathrm{d}x = \sec x + c$$

$$\int \operatorname{cosec} x \cot x\mathrm{d}x = -\operatorname{cosec} x + c$$

$$\int \operatorname{cosec}^2 x\mathrm{d}x = -\cot x + c$$

Examples

(a) $\int 2\sec 3x \tan 3x\mathrm{d}x = \tfrac{2}{3}\sec 3x + c$

Look out for the function written in an alternative way.

(b) $\int \frac{1}{\sin^2 2x}\mathrm{d}x = \int \operatorname{cosec}^2 2x\mathrm{d}x = -\tfrac{1}{2}\cot 2x + c$

Integrals leading to a logarithmic function

AQA	P4
EDEXCEL	P3
OCR	P3
WJEC	P3
NICCEA	P3

Remember that $\frac{d}{dx}\ln(f(x)) = \frac{f'(x)}{f(x)}$.

Applying the process in reverse gives:

$$\int \frac{f'(x)}{f(x)}\,dx = \ln(f(x)) + c$$

KEY POINT

Examples

(a) $\int \frac{x^2}{2x^3 - 2}\,dx = \frac{1}{6}\ln(2x^3 - 2) + c$

$\frac{d}{dx}(2x^3 - 2) = 6x^2$ so a factor of $\frac{1}{6}$ is needed.

(b) $\int \tan x\,dx = \int \frac{\sin x}{\cos x}\,dx$

$= -\ln(\cos x) + c$

$= \ln(\sec x) + c$

$-\ln a = \ln(a^{-1})$

Key points from AS

- **Modulus function** *Revise AS page 50*
- **Area under curve** *Revise AS page 42*

Care must be taken with **definite integrals** of this type.
These can be evaluated between the limits a and b only if $f(x)$ exists for all values between a and b. Note that $f(a)$ and $f(b)$ will have the same sign.

$$\int_a^b \frac{f'(x)}{f(x)}\,dx = \Big[\ln|f(x)|\Big]_a^b = (\ln|f(b)| - \ln|f(a)|)$$

Example

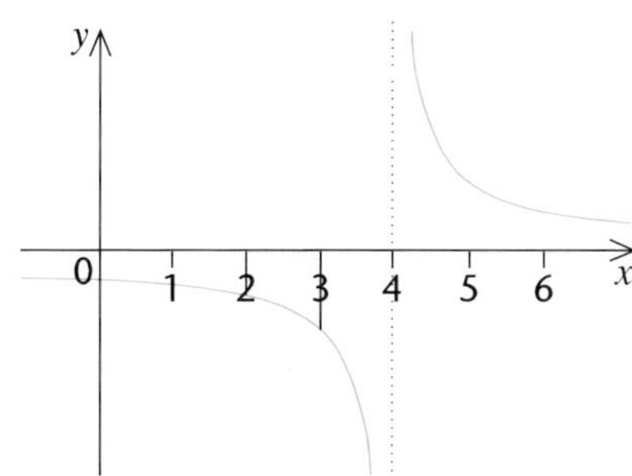

The diagram shows the curve $y = \frac{1}{x-4}$.

Either: find the shaded area
or: find the area between the curve, the x-axis, and the lines $x = 2$ and $x = 3$.

$\int_a^b y\,dx = \int_2^3 \frac{1}{x-4}\,dx$

$= \Big[\ln|x-4|\Big]_2^3$

$= \ln|-1| - \ln|-2|$

$= \ln 1 - \ln 2$

$= -\ln 2 (= -0.69 \ldots)$

$\ln|-2| = \ln 2$
$\ln 1 = 0$

Note that the formula gives a negative value, confirming that the area is below the x-axis.

Shaded area $= \ln 2 (= 0.69 \ldots)$ square units.

Note that $\int_3^5 \frac{1}{x-4}\,dx$ cannot be evaluated, as the curve is undefined at $x = 4$.

$f(3)$ and $f(5)$ do not have the same sign.

Partial fractions

AQA	P4
EDEXCEL	P3
OCR	P3
WJEC	P3
NICCEA	P3

Some rational functions can be integrated once they have been expressed in partial fraction form.

Decompose into partial fractions (see p. 19).

Key points from AS

- **Logarithms**
  Revise AS page 55

Example

Find $\int \frac{x+7}{(x-2)(x+1)}\,dx$, simplifying your answer.

$$\frac{x+7}{(x-2)(x+1)} \equiv \frac{3}{x-2} - \frac{2}{x+1}$$

$$\Rightarrow \int \frac{x+7}{(x-2)(x+1)}\,dx = \int \left(\frac{3}{x-2} - \frac{2}{x+1}\right)dx$$

$$= 3\ln(x-2) - 2\ln(x+1) + c$$

$$= \ln\frac{(x-2)^3}{(x+1)^2} + c$$

The working for this example is shown on p.19

$\log a - \log b = \log\left(\frac{a}{b}\right)$

$\log a^n = n\log a$

Substitution

AQA	P5
EDEXCEL	P3
OCR	P3
WJEC	P2
NICCEA	P3

Sometimes an integral is made easier by using a substitution.

This transforms the integral with respect to one variable, say x, into an integral with respect to a related variable, say u.

This method is known as the method of **substitution** or **change of variable**.

KEY POINT

$$\int f(x)\,dx = \int f(x)\frac{dx}{du}\,du$$

This could also be done by using the substitution $u = 2x + 1$.

Example

Find $\int x\sqrt{2x+1}\,dx$, using the substitution $u = \sqrt{2x+1}$.

$$\int x\sqrt{2x+1}\,dx = \int x\sqrt{2x+1}\frac{dx}{du}\,du$$

$$= \int \tfrac{1}{2}(u^2-1)u \times u\,du$$

$$= \frac{1}{2}\int (u^4 - u^2)\,du$$

$$= \frac{1}{2}\left(\frac{u^5}{5} - \frac{u^3}{3}\right) + c$$

$$= \frac{u^3}{2}\left(\frac{u^2}{5} - \frac{1}{3}\right) + c$$

$$= \tfrac{1}{30}u^3(3u^2 - 5) + c$$

$$= \tfrac{1}{30}(2x+1)^{\frac{3}{2}}(6x-2) + c$$

$$= \tfrac{1}{15}(2x+1)^{\frac{3}{2}}(3x-1) + c$$

Side working

$u = \sqrt{2x+1}$

$u^2 = 2x + 1 \Rightarrow x = \tfrac{1}{2}(u^2 - 1)$

$\frac{dx}{du} = u$

$3u^2 - 5 = 3(2x+1) - 5$
$= 6x - 2$
$= 2(3x - 1)$

Show the side working as part of your answer.

When simplifying, take out factors as soon as possible.

Write the answer in terms of x, simplifying as much as possible.

When evaluating definite integrals using the method of substitution, change the x limits to u limits and substitute them into the working as soon as possible. There is no need to tidy up the expression first.

Sometimes it is possible to bypass the method of substitution by **recognising** an integral as the derivative of a particular function.

This type of integral can always be done by using a substitution.

Examples

(a) $\int (2x+1)^5 dx = \frac{1}{12}(2x+1)^6 + c$

Recognise that $\frac{d}{dx}(2x+1)^6 = 12(2x+1)^5$

(b) $\int x\sqrt{2x^2+1}\, dx = \frac{1}{6}(2x^2+1)^{\frac{3}{2}} + c$

Recognise that $\frac{d}{dx}(2x^2+1)^{\frac{3}{2}} = 6x(2x^2+1)^{\frac{1}{2}}$

General result:

$$\int f'(x)[f(x)]^n dx = \frac{1}{n+1}[f(x)]^{n+1} + c$$

Integration by parts

AQA	P5
EDEXCEL	P3
OCR	P3
WJEC	P3
NICCEA	P3

It is sometimes possible to integrate a product using **integration by parts**.

KEY POINT

If u and v are functions of x, then $\int u\frac{dv}{dx}dx = uv - \int v\frac{du}{dx}dx$

Before using integration by parts, check whether the function can be simplified, or whether the integral can be recognised directly or done using a substitution.

Example

Find $\int 2x \sin x dx$.

Let $u = 2x$, then $\frac{du}{dx} = 2$

Let $\frac{dv}{dx} = \sin x$ then $v = -\cos x$

Integration by parts can only be used if v can be found from $\frac{dv}{dx}$ and it is possible to find $\int v\frac{du}{dx}dx$.

$$\therefore \int 2x \sin x dx = 2x(-\cos x) - \int(-\cos x \times 2)dx$$
$$= -2x\cos x + \int 2\cos x dx$$
$$= -2x\cos x + 2\sin x + c$$

Sometimes the integration by parts process has to be carried out more than once. This happens when integrating expressions such as x^2e^x or $x^2 \cos x$.

Definite integrals

$$\int_a^b u\frac{dv}{dx}dx = \Big[uv\Big]_a^b - \int_a^b v\frac{du}{dx}dx$$

Example

$$\int_0^1 xe^{2x}dx = \left[x\frac{1}{2}e^{2x}\right]_0^1 - \int_0^1 \frac{1}{2}e^{2x}dx$$
$$= \tfrac{1}{2}e^2 - \left[\tfrac{1}{4}e^{2x}\right]_0^1$$
$$= \tfrac{1}{2}e^2 - (\tfrac{1}{4}e^2 - \tfrac{1}{4}) = \tfrac{1}{4}e^2 + \tfrac{1}{4}$$

Often a polynomial function is taken as u as it becomes of lower degree when differentiated. Integrals involving $\ln x$ are an exception, since $\ln x$ is not easy to integrate, but can be differentiated.

Some functions containing $\ln x$ can be integrated using integration by parts. A useful strategy is to take $\ln x$ as u.

Example

$$\int x^2 \ln x\,dx = \int (\ln x \times x^2)\,dx$$
$$= \ln x \times \left(\frac{x^3}{3}\right) - \int \left(\frac{x^3}{3} \times \frac{1}{x}\right)dx$$
$$= \tfrac{1}{3}x^3 \ln x - \frac{1}{3}\int x^2\,dx$$
$$= \tfrac{1}{3}x^3 \ln x - \tfrac{1}{9}x^3 + c$$
$$= \tfrac{1}{9}x^3(3 \ln x - 1) + c$$

If $u = \ln x$, then $\frac{du}{dx} = \frac{1}{x}$

This method is particularly useful when integrating $\ln x$.
Think of it as integrating the product $\ln x \times 1$ as follows:

$$\int (\ln x \times 1)\,dx = \ln x \times x - \int \left(x \times \frac{1}{x}\right)dx$$
$$= x \ln x - \int 1\,dx$$
$$= x \ln x - x + c$$

Parametric integration

AQA P5
EDEXCEL P3

When a curve is defined in terms of a parameter such that $x = f(t)$ and $y = g(t)$ then the formula for the area under a curve can be adapted as follows:

$$A = \int_{x_1}^{x_2} y\,dx = \int_{t_1}^{t_2} g(t)\frac{dx}{dt}\,dt$$

KEY POINT

Example
Using parametric integration, find the area of the ellipse defined by $x = 2 \cos \theta$, $y = 3 \sin \theta$ $(0 \leqslant \theta \leqslant 2\pi)$.

The sign of the area given by the formula depends on the direction in which the curve is being traced out as the parameter increases: positive for clockwise, negative for anticlockwise.

In general, area of ellipse defined by $x = a \cos \theta$, $y = b \sin \theta$ is πab.

$$A = \int_0^{2\pi} y\frac{dx}{d\theta}\,d\theta$$
$$= \int_0^{2\pi} 3 \sin \theta(-2 \sin \theta)\,d\theta$$
$$= -6\int_0^{2\pi} \sin^2 \theta\,d\theta$$
$$= -6\int_0^{2\pi} \tfrac{1}{2}(1 - \cos 2\theta)\,d\theta$$
$$= -3\left[\theta - \tfrac{1}{2}\sin 2\theta\right]_0^{2\pi}$$
$$= -3(2\pi - 0 - (0 - 0))$$
$$= -6\pi$$

This curve is traced out in an anticlockwise direction

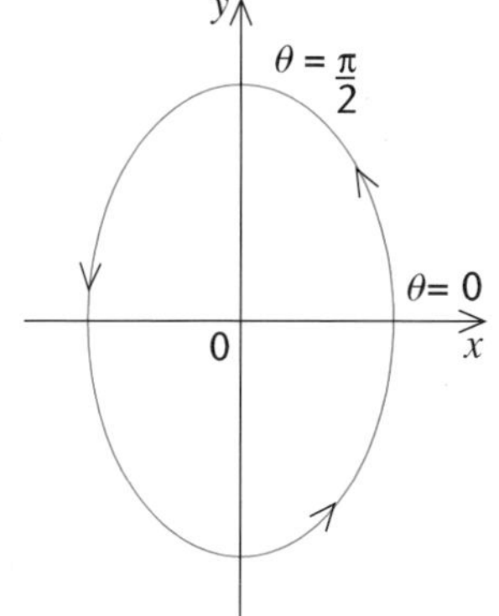

The area of the ellipse is 6π square units.

Differential equations

AQA	P5
EDEXCEL	P3
OCR	P3
WJEC	P3
NICCEA	P3

A **first order differential equation** in x and y contains only the first differential coefficient.

Examples $\frac{dy}{dx} = -2x$, $\frac{dy}{dx} = \frac{x}{y}$, $y\frac{dy}{dx} = 3$

If $\frac{dy}{dx} = -2x$, then integrating with respect to x gives $y = -x^2 + c$.

The general solution contains an arbitrary integration constant.

This is the **general solution** of the differential equation and it can be illustrated by a family of curves.

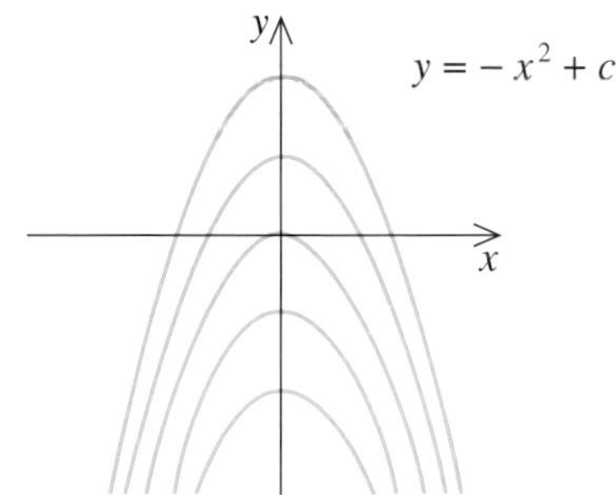

A **particular solution** is a specific member of the family and it can be found from knowledge of additional information.

The value of c is found to give a particular solution.

For example, if the curve goes through the point (2, 0), then $0 = -(2)^2 + c \Rightarrow c = 4$. The particular solution is $y = -x^2 + 4$.

Separating the variables

> **KEY POINT**
>
> A differential equation that can be written in the form $f(y)\frac{dy}{dx} = g(x)$ can be solved by **separating the variables, where** $\int f(y)dy = \int g(x)dx$.

Example

Solve (a) $\frac{dy}{dx} = \frac{x}{y}$ (b) $\frac{dy}{dx} = y$

(a) $\frac{dy}{dx} = \frac{x}{y}$

$$\int y\,dy = \int x\,dx \Rightarrow \tfrac{1}{2}y^2 = \tfrac{1}{2}x^2 + c$$

Separate the variables to get the format $\int \ldots dy = \int \ldots dx$.

(b) $\frac{dy}{dx} = y$

$$\frac{1}{y}\frac{dy}{dx} = 1$$

This is a special case, when $g(x) = 1$.

$$\int \frac{1}{y}dy = \int 1dx$$

$$\ln y = x + c$$

The answer can be left like this.

$$y = e^{x+c} = e^x \times e^c = Ae^x \quad (\text{where } A = e^c)$$

So $\frac{dy}{dx} = y \Rightarrow y = Ae^x$

The letter A is often used here, but it could be any letter (other than e, x or y, of course!).

Exponential growth and decay

AQA	P4
EDEXCEL	P3
OCR	P2
WJEC	P3

You may have to form a differential equation from given information. Often this models exponential growth or exponential decay.

Exponential growth

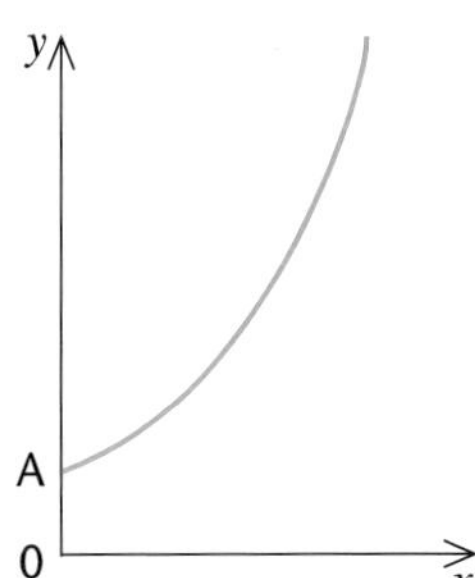

$$\frac{dy}{dx} = ky \Rightarrow y = Ae^{kx} \quad (k > 0)$$

Exponential decay

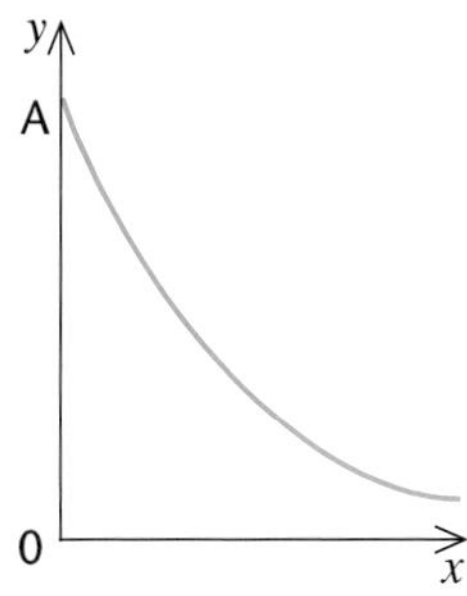

$$\frac{dy}{dx} = -ky \Rightarrow y = Ae^{-kx} \quad (k > 0)$$

In exponential decay, the fact that y is decreasing with respect to x is usually shown by the negative, keeping $k > 0$.

An example of exponential growth is the growth of a population where $\frac{dp}{dt} = kt \Rightarrow p = p_0 e^{-kt}$

p_0 is the initial population.

Examples of exponential decay are:

(a) the disintegration of radioactive materials, $\frac{dm}{dt} = -km \Rightarrow m = m_0 e^{-kt}$

m_0 is the initial mass.

(b) Newton's Law of Cooling, $\frac{dT}{dt} = -k(T - T_s) \Rightarrow T - T_s = (T_0 - T_s)e^{-kt}$

T_s is temperature of surroundings, T_0 is initial temperature of the object.

Progress check

1 (a) $\int 4 \sin 2x\,dx$ (b) $\int \sec^2 2x\,dx$ (c) $\int \cos (\tfrac{1}{2}x)dx$

(d) $\int 4 \sin^2 x\,dx$ (e) $\int \cos^2 2x\,dx$ (f) Evaluate $\int_0^{\frac{1}{6}\pi} \sin^4 x \cos x\,dx$

2 Find (a) $\int \frac{1}{2x+5}\,dx$ (b) $\int \cot x\,dx$ (c) $\int \frac{3x+2}{x^2-x-12}\,dx$

Split (c) into partial fractions first.

3 Using the substitution $u = 2 + e^{2x}$, or otherwise, find $\int \frac{e^{2x}}{2+e^{2x}}\,dx$

Remember that $\frac{dx}{du} = \frac{1}{du/dx}$.

4 (a) Find (i) $\int xe^{-x}dx$ (ii) $\int x^3 \ln x\,dx$ (b) Evaluate $\int_0^{\frac{1}{2}\pi} x \cos x\,dx$

5 Solve the differential equation $\frac{dy}{dx} = 2xy$, given that $y = 3$ when $x = 0$.

1 (a) $-2\cos 2x + c$ (b) $\frac{1}{2}\tan 2x + c$ (c) $2\sin(\frac{1}{2}x) + c$
(d) $2x - \sin 2x + c$ (e) $\frac{1}{2}x + \frac{1}{8}\sin 4x + c$ (f) $\frac{1}{160}$
2 (a) $\frac{1}{2}\ln(2x+5) + c$ (b) $\ln(\sin x) + c$ (c) $\ln(x+3)(x-4)^2 + c$
3 $\frac{1}{2}\ln(2+e^{2x}) + c$
4 (a) (i) $-e^{-x}(x+1) + c$ (ii) $\frac{1}{16}x^4(4\ln x - 1) + c$ (b) $\frac{1}{2}\pi - 1$
5 $y = 3e^{x^2}$

1.5 Numerical methods

After studying this section you should be able to:

- *find an approximate solution to an equation using the Newton-Raphson method*
- *perform numerical integration using the mid-ordinate rule and Simpson's rule*

LEARNING SUMMARY

Newton-Raphson method

AQA P4
NICCEA P3

The **Newton-Raphson method** can be used to find an approximate solution to an equation.

Consider the equation $f(x) = 0$, with unknown root α, i.e. $f(\alpha) = 0$.

This process can then be repeated several times to obtain a more accurate value for the solution.

If x_n is a good approximation to α, then a better approximation* is given by

$$x_{n+1} = x_n - \frac{f(x_n)}{f'(x_n)}.$$

* There are cases when this method fails to converge, for example when $f'(\alpha)$ is near to zero or when $f'(\alpha)$ is very large. There may also be problems if there are discontinuities in the curve.

Example

The equation $f(x) = 2\cos x - x^2$ has root α.

(a) Show that α lies between 1 and 1.1.
(b) Using 1 as a first approximation, find α correct to 2 decimal places.

Remember to work in radians.

$$f(x) = 2\cos x - x^2$$
$$f(1) = 2\cos 1 - 1 = 0.0806$$
$$f(1.1) = 2\cos 1.1 - 1.1^2 = -0.3028\ \ldots$$

Key points from AS

- **Sign change**
  Revise AS page 64

Since $f(1) > 0$ and $f(1.1) < 0$, there must be a value between $x = 1$ and $x = 1.1$ for which $f(x) = 0$. Therefore α lies between 1 and 1.1.

$$f'(x) = -2\sin x - 2x$$

Taking $x_1 = 1$,

$$x_2 = 1 - \frac{f(1)}{f'(1)}$$
$$= 1 - \frac{2\cos 1 - 1}{-2\sin 1 - 2}$$
$$= 1 - \frac{0.0806\ \ldots}{-3.6829\ \ldots}$$
$$= 1.021885\ \ldots$$

Taking $x_2 = 1.02189$,

$$x_3 = 1.02189 - \frac{f(1.02189)}{f'(1.02189)}$$
$$= 1.02189 - \frac{-0.000750\ \ldots}{-3.7499\ \ldots}$$
$$= 1.02169\ \ldots$$

The approximations agree to 2 d.p.

$$\therefore \quad \alpha = 1.02 \text{ (2 d.p.)}$$

Numerical integration

AQA P5

Key points from AS

- **Area under curve** *Revise AS page 42*
- **The trapezium rule** *Revise AS page 66*

Consider the curve $y = f(x)$.

The area enclosed by the curve, the x-axis, $x = a$ and $x = b$ is found by evaluating $\int_a^b f(x)\mathrm{d}x$.

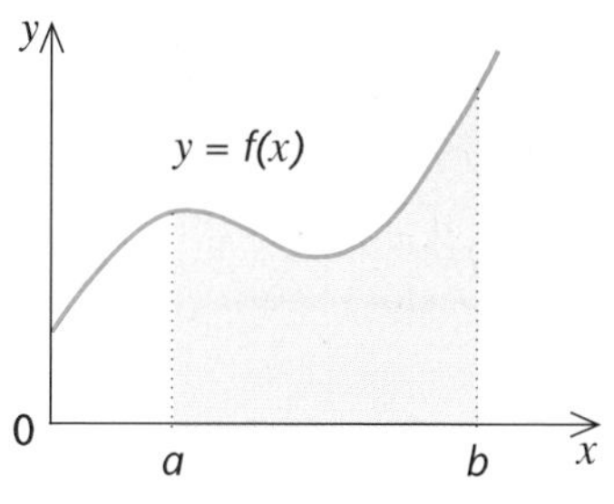

There are several ways of obtaining an approximation to this value, which are useful when it is difficult or not possible to evaluate the integral.
Methods include the trapezium rule, **Simpson's rule** and the **mid-ordinate rule**.

Simpson's rule

This method divides the area into an **even number** of parallel strips and approximates the areas of pairs of strips using the following formula:

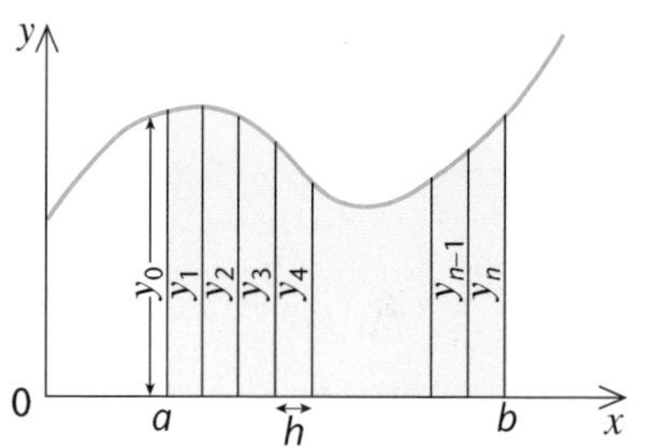

$$\int_a^b f(x)\mathrm{d}x \approx \tfrac{1}{3}h\{(y_0 + y_n) + 4(y_1 + y_3 + \ldots + y_{n-1}) + 2(y_2 + y_4 + \ldots + y_{n-2})\}$$

where $h = \dfrac{b-a}{n}$ and n is even.

KEY POINT

Note that the first ordinate has been labelled y_0 and the last y_n. There are n strips and $n + 1$ ordinates.

Example

Estimate $\int_0^1 e^{x^2}\mathrm{d}x$ using Simpson's rule with 10 strips.

Tabulating the results helps in doing the final calculation.

If you have multiple memories in your calculator, retain the figures. Otherwise approximate say to 4 decimal places.

x	y	*First and last ordinates*	*Odd ordinates*	*Other ordinates*
0	y_0	1		
0.1	y_1		1.010 ...	
0.2	y_2			1.040 ...
0.3	y_3		1.094 ...	
0.4	y_4			1.173 ...
0.5	y_5		1.284 ...	
0.6	y_6			1.433 ...
0.7	y_7		1.632 ...	
0.8	y_8			1.896 ...
0.9	y_9		2.247 ...	
1	y_{10}	2.718 ...		
Totals		3.718 ...	7.268 ...	5.544 ...

$h = \dfrac{1-0}{10} = 0.1$

$$\int_a^b f(x)\mathrm{d}x \approx \tfrac{1}{3}h\{(y_0 + y_{10}) + 4(y_1 + y_3 + y_5 + y_7 + y_9) + 2(y_2 + y_4 + y_6 + y_8)\}$$

$= \tfrac{1}{3} \times 0.1 \times \{3.718 \ldots + 4(7.268 \ldots) + 2(5.544 \ldots)\}$

$= 1.463$ (3 s.f.)

The mid-ordinate rule

The more rectangles that are taken, the better the approximation.

The area is split into rectangles of equal width h.
The height of each rectangle is taken as the y-value halfway along each rectangle.

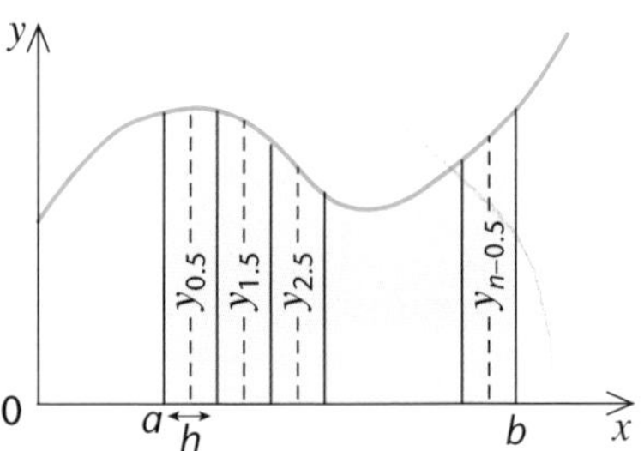

Denoting the mid-ordinates by $y_{0.5}, y_{1.5}, \dots, y_{n-0.5}$, then

KEY POINT

$$\int_a^b f(x)\mathrm{d}x \approx h(y_{0.5} + y_{1.5} + \dots y_{n\;0.5}) \text{ where } h = \frac{b-a}{n}$$

Example

Estimate $\int_1^3 \ln x\mathrm{d}x$ using the mid-ordinate rule with 4 rectangles.

There are 4 rectangles, so $h = \frac{3-1}{4} = 0.5$,

$x_0 = 1,\ x_1 = 1.5,\ x_2 = 2,\ x_3 = 2.5,\ x_4 = 3.$

$\therefore\quad x_{0.5} = 1.25 \Rightarrow y_{0.5} = \ln 1.25.$

Calculating the other y mid-ordinates in a similar way gives:

$$\int_1^3 \ln x\mathrm{d}x \approx 0.5(\ln 1.25 + \ln 1.75 + \ln 2.25 + \ln 2.75)$$

$$= 1.3026 \dots = 1.30 \text{ (2 d.p.)}$$

Using integration by parts, the exact answer is $3 \ln 3 - 2 = 1.295 \dots$ so the answers agree to 2 d.p.

Progress check

1 Show that the equation $e^x + 2x^2 - 2 = 0$ has a root between $x = 0.4$ and $x = 0.5$. Using the Newton-Raphson method with first approximation 0.5, find the value of the root correct to 2 decimal places.

2 Estimate $\int_0^1 \sqrt{1 - x^3}\ \mathrm{d}x$

(a) using Simpson's rule with 4 strips
(b) using the mid-ordinate rule with 4 rectangles
(c) using the trapezium rule with 4 trapezia.

1 0.46
2 To 3 d.p. (a) 0.823 (b) 0.854 (c) 0.797

1.6 Vectors

After studying this section you should be able to:

LEARNING SUMMARY

- *use unit vectors, position vectors and displacement vectors, distance between two points*
- *use scalar products to find the angle between two directions*
- *express equations of lines in vector form*
- *determine whether two lines are parallel, intersect or are skew*
- *find the equation of a plane and the angle between two planes*

Scalar product

AQA	P5
EDEXCEL	P3
OCR	P3
WJEC	P3
NICCEA	P3

If $\mathbf{a} = x_1\mathbf{i} + y_1\mathbf{j} + z_1\mathbf{k}$, then $|\mathbf{a}| = \sqrt{x_1^2 + y_1^2 + z_1^2}$ where $|\mathbf{a}|$ is the length of $\mathbf{a}$.

If $\mathbf{b} = x_2\mathbf{i} + y_2\mathbf{j} + z_2\mathbf{k}$, then $|\mathbf{b}| = \sqrt{x_2^2 + y_2^2 + z_2^2}$.

Key points from AS

Vectors
Revise AS pages 72–74

KEY POINT

If θ is the angle between $\mathbf{a}$ and $\mathbf{b}$, the **scalar** (or **dot**) **product** of $\mathbf{a}$ and $\mathbf{b}$ is $\mathbf{a.b}$ where $\mathbf{a.b} = |\mathbf{a}||\mathbf{b}|\cos\theta$.

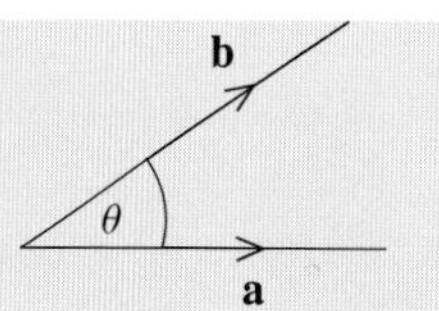

This leads to the useful results:
$\mathbf{i.i} = 1$, $\mathbf{j.j} = 1$, $\mathbf{k.k} = 1$
$\mathbf{i.j} = \mathbf{j.k} = \mathbf{k.i} = 0$

If two vectors are **parallel**, $\theta = 0$ and $\mathbf{a.b} = |\mathbf{a}||\mathbf{b}|$.
If two vectors are **perpendicular**, $\theta = 90°$ and $\mathbf{a.b} = 0$.

For $\mathbf{a}$ and $\mathbf{b}$ defined as above:

In column vectors

$$\mathbf{a.b} = \begin{pmatrix} x_1 \\ y_1 \\ z_1 \end{pmatrix} \cdot \begin{pmatrix} x_2 \\ y_2 \\ z_2 \end{pmatrix} = x_1x_2 + y_1y_2 + z_1z_2$$

$$\mathbf{a.b} = (x_1\mathbf{i} + y_1\mathbf{j} + z_1\mathbf{k}).(x_2\mathbf{i} + y_2\mathbf{j} + z_2\mathbf{k})$$
$$= x_1x_2 + y_1y_2 + z_1z_2$$

This uses the results for parallel and perpendicular vectors.

Example

Find the angle between $\mathbf{a} = 2\mathbf{i} + \mathbf{j} + 4\mathbf{k}$ and $\mathbf{b} = -3\mathbf{i} + 2\mathbf{j} - \mathbf{k}$.

$$\mathbf{a.b} = 2(-3) + 1(2) + 4(-1) = -8$$
$$|\mathbf{a}| = \sqrt{2^2 + 1^2 + 4^2} = \sqrt{21}$$
$$|\mathbf{b}| = \sqrt{(-3)^2 + 2^2 + (-1)^2} = \sqrt{14}$$
$$\mathbf{a.b} = |\mathbf{a}||\mathbf{b}|\cos\theta$$
$$-8 = \sqrt{21}\,\sqrt{14}\cos\theta$$
$$\Rightarrow \cos\theta = \frac{-8}{\sqrt{21}\,\sqrt{14}} = -0.4665\ldots$$
$$\Rightarrow \theta = 117.8° \text{ (1 d.p.)}$$

In column vectors:

$$\mathbf{a.b} = \begin{pmatrix} 2 \\ 1 \\ 4 \end{pmatrix} \cdot \begin{pmatrix} -3 \\ 2 \\ -1 \end{pmatrix} = -8$$

Vector equation of a line

AQA	P5
EDEXCEL	P3
OCR	P3
WJEC	P3
NICCEA	P3

KEY POINT

A **vector equation** of a line passing through a fixed point A with position vector $\mathbf{a}$ and parallel to a vector $\mathbf{b}$ is $\mathbf{r} = \mathbf{a} + t\mathbf{b}$, where t is a scalar parameter.

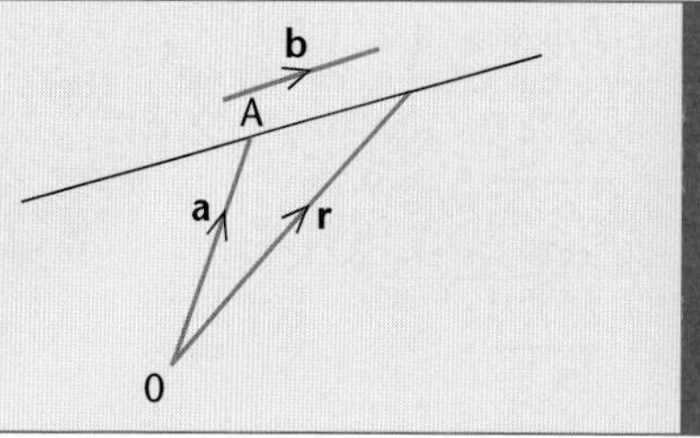

t is often used for the parameter, but not exclusivley so. Other letters sometimes used include s, μ and λ.

$\mathbf{r}$ is the position vector of any point on the line.
$\mathbf{b}$ is the **direction vector** of the line.

The equation of the line is not unique. Any other point on the line could be used instead of A and any multiple of $\mathbf{b}$ could be used for the direction vector.

In column format: $\mathbf{r} = \begin{pmatrix} 4 \\ -5 \\ 1 \end{pmatrix} + t\begin{pmatrix} 3 \\ 4 \\ -2 \end{pmatrix}$.

Example

A vector equation of the line passing through (4, –5, 1), parallel to $3\mathbf{i} + 4\mathbf{j} - 2\mathbf{k}$ is $\mathbf{r} = 4\mathbf{i} - 5\mathbf{j} + \mathbf{k} + t(3\mathbf{i} + 4\mathbf{j} - 2\mathbf{k})$.

KEY POINT

A vector equation of the line through two fixed points A and B, with position vectors $\mathbf{a}$ and $\mathbf{b}$ is given by $\mathbf{r} = \mathbf{a} + t(\mathbf{b} - \mathbf{a})$ where t is a scalar parameter.

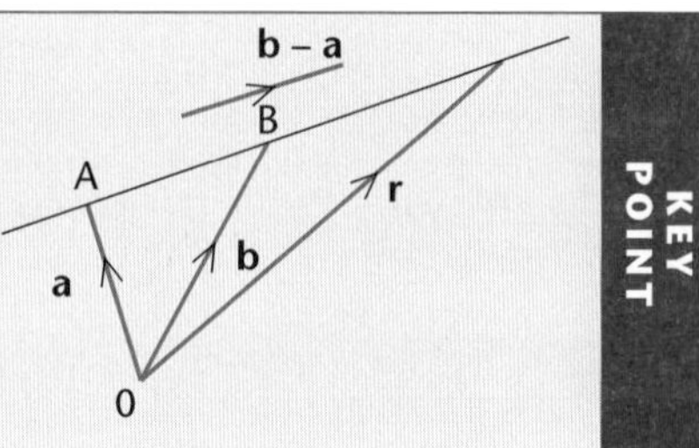

Example

Find a vector equation of the line through the points $P(3, 1, -4)$ and $Q(-2, 5, 1)$.

Direction of line PQ:

$\overrightarrow{PQ} = \mathbf{q} - \mathbf{p} = -2\mathbf{i} + 5\mathbf{j} + \mathbf{k} - (3\mathbf{i} + \mathbf{j} - 4\mathbf{k}) = -5\mathbf{i} + 4\mathbf{j} + 5\mathbf{k}$

A vector equation of PQ is $\mathbf{r} = 3\mathbf{i} + \mathbf{j} - 4\mathbf{k} + t(-5\mathbf{i} + 4\mathbf{j} + 5\mathbf{k})$.

Angles between two lines

To find the angle between two lines, find the angle between their direction vectors.

Pairs of lines

AQA	P5
OCR	P3
WJEC	P3

In 2-dimensions, a pair of lines are either parallel or they intersect.
In 3-dimensions, a pair of lines are parallel, or they intersect or they are **skew**.

If two lines are parallel, then their direction vectors are multiples of each other, as in this **example**:
$\mathbf{r} = 3\mathbf{i} - 5\mathbf{j} + 2\mathbf{k} + \lambda(\mathbf{i} + 2\mathbf{j} - \mathbf{k})$ and $\mathbf{r} = 4\mathbf{i} + \mathbf{j} - 2\mathbf{k} + \mu(3\mathbf{i} + 6\mathbf{j} - 3\mathbf{k})$.

The two lines $\mathbf{r} = \mathbf{a} + \lambda\mathbf{b}$ and $\mathbf{r} = \mathbf{c} + \mu\mathbf{d}$ intersect if unique values of λ and μ can be found such that $\mathbf{a} + \lambda\mathbf{b} = \mathbf{c} + \mu\mathbf{d}$.
If unique values cannot be found, then the two lines are **skew**.

If the lines intersect, then $\mathbf{r}_1 = \mathbf{r}_2$ will have a unique solution for λ and μ.

Note that the lines are not parallel, since their direction vectors are not multiples of each other.

Example

Investigate whether the lines $\mathbf{r}_1 = 2\mathbf{i} + \mathbf{j} - 3\mathbf{k} + \lambda(4\mathbf{i} + 6\mathbf{j} - \mathbf{k})$ and $\mathbf{r}_2 = 3\mathbf{i} - 2\mathbf{k} + \mu(4\mathbf{i} + \mathbf{j} + 3\mathbf{k})$ intersect or whether they are skew.

$\mathbf{r}_1 = \mathbf{r}_2 \Rightarrow 2\mathbf{i} + \mathbf{j} - 3\mathbf{k} + \lambda(4\mathbf{i} + 6\mathbf{j} - \mathbf{k}) = 3\mathbf{i} - 2\mathbf{k} + \mu(4\mathbf{i} + \mathbf{j} + 3\mathbf{k})$

i.e. $(2 + 4\lambda)\mathbf{i} + (1 + 6\lambda)\mathbf{j} + (-3 - \lambda)\mathbf{k} = (3 + 4\mu)\mathbf{i} + (0 + \mu)\mathbf{j} + (-2 + 3\mu)\mathbf{k}$

Equating coefficients:

$2 + 4\lambda = 3 + 4\mu \quad (1)$
$1 + 6\lambda = \mu \quad (2)$
$-3 - \lambda = -2 + 3\mu \quad (3)$

Solving (1) and (2) simultaneously gives:
$\lambda = -0.25$ and $\mu = -0.5$
Check whether these satisfy (3):
LHS $= -3 - (-0.25) = -2.75$
RHS $= -2 + 3(-0.5) = -3.5$

$\therefore \lambda = -0.25$ and $\mu = -0.5$ do not satisfy all three equations.
Since there is not a unique solution for λ and μ, the lines do not intersect. They are skew.

Planes

AQA P5

> A vector equation of a plane through a point with position vector $\mathbf{a}$, perpendicular to a vector $\mathbf{n}$ is given by $(\mathbf{r}-\mathbf{a}).\mathbf{n}=0 \Rightarrow \mathbf{r}.\mathbf{n}=\mathbf{a}.\mathbf{n}$
>
> KEY POINT

Example

Find a vector equation of a plane through the point with position vector $\mathbf{i}+4\mathbf{j}+5\mathbf{k}$, perpendicular to the vector $-2\mathbf{i}+3\mathbf{j}+\mathbf{k}$. Also give the equation in Cartesian form.

A vector equation of the plane is $\mathbf{r}.\mathbf{n}=\mathbf{a}.\mathbf{n}$

$\Rightarrow \mathbf{r}.(-2\mathbf{i}+3\mathbf{j}+\mathbf{k})=(\mathbf{i}+4\mathbf{j}+5\mathbf{k}).(-2\mathbf{i}+3\mathbf{j}+\mathbf{k})$

$\Rightarrow \mathbf{r}.(-2\mathbf{i}+3\mathbf{j}+\mathbf{k})=15$

$$\begin{pmatrix}1\\4\\5\end{pmatrix}\cdot\begin{pmatrix}-2\\3\\1\end{pmatrix}$$
$=1(-2)+4(3)+5(1)=15$

In Cartesian form, a general point in the plane is $\mathbf{r}=x\mathbf{i}+y\mathbf{j}+z\mathbf{k}$,
so $\mathbf{r}.\mathbf{n}=(x\mathbf{i}+y\mathbf{j}+z\mathbf{k}).(-2\mathbf{i}+3\mathbf{j}+\mathbf{k})=-2x+3y+z$

Note that −2, 3 and 1 are the components of the normal vector to the plane.

But from above, $\mathbf{r}.\mathbf{n}=15$,
so the equation of the plane is $-2x+3y+z=15$.

The angle between two planes is the angle between the normal vectors to the planes, where $\mathbf{n}_1.\mathbf{n}_2=|\mathbf{n}_1||\mathbf{n}_2|\cos\theta$.

Progress check

1 Find the angle between the vectors $\mathbf{a}=2\mathbf{i}+3\mathbf{j}-\mathbf{k}$ and $\mathbf{b}=\mathbf{i}+\mathbf{j}+2\mathbf{k}$.

2 (a) Give a vector equation of the line which is parallel to $4\mathbf{i}-\mathbf{j}+2\mathbf{k}$ and goes through the point with coordinates (3, 0, 1).
(b) Give a vector equation of the line through (4, −1, 1) and (−3, 2, −2).

3 (a) Find the point of intersection of the lines

$\mathbf{r}=2\mathbf{i}-3\mathbf{j}+\mu(\mathbf{i}-2\mathbf{j})$ and $\mathbf{r}=\mathbf{i}-2\mathbf{j}+\lambda(2\mathbf{i}+3\mathbf{j})$.

(b) Investigate whether the following lines intersect or whether they are skew. If they intersect, find their point of intersection.

$\mathbf{r}=\mathbf{i}+2\mathbf{j}-5\mathbf{k}+s(3\mathbf{i}+\mathbf{j}+\mathbf{k})$ and $\mathbf{r}=2\mathbf{i}-\mathbf{j}+2\mathbf{k}+t(-\mathbf{i}-\mathbf{j}+\mathbf{k})$

(c) Find the angle between the lines given in (b).

4 Find the angle between the two planes

$3x+y-2z=10$ and $2x-y+2z=15$

1 70.9°
2 Answers are not unique
(a) $\mathbf{r}=3\mathbf{i}+\mathbf{k}+t(4\mathbf{i}-\mathbf{j}+2\mathbf{k})$ (b) $\mathbf{r}=4\mathbf{i}-\mathbf{j}+\mathbf{k}+t(-7\mathbf{i}+3\mathbf{j}-3\mathbf{k})$
3 (a) $(1\frac{2}{7}, -1\frac{4}{7})$ (b) Intersect at (7, 4, −3) (c) 121.5°
4 84.9°

Sample questions and model answers

1

(a) Express $f(x) = \dfrac{5x+1}{(x+2)(x-1)^2}$ in partial fractions.

(b) Hence show that $\int_{-1}^{0} f(x)\mathrm{d}x = 1 - 2\ln 2$.

Remember the format when there is a repeated factor.

Take care with the denominator.

(a) Let $\dfrac{5x+1}{(x+2)(x-1)^2} \equiv \dfrac{A}{x+2} + \dfrac{B}{x-1} + \dfrac{C}{(x-1)^2}$

$$\Rightarrow \frac{5x+1}{(x+2)(x-1)^2} \equiv \frac{A(x-1)^2 + B(x+2)(x-1) + C(x+2)}{(x+2)(x-1)^2}$$

$$\Rightarrow 5x+1 \equiv A(x-1)^2 + B(x+2)(x-1) + C(x+2)$$

Putting $x = 1$ makes the factor $(x-1)$ equal to zero.

You could equate the x term or the constant term if you wish.

Let $x = 1$: $6 = 3C \Rightarrow C = 2$

Let $x = -2$: $-9 = 9A \Rightarrow A = -1$

Equate x^2 terms: $0 = A + B \Rightarrow B = 1$

$$\therefore\ f(x) = -\frac{1}{x+2} + \frac{1}{x-1} + \frac{2}{(x-1)^2}$$

Remember the modulus sign.

$\ln 1 = \ln|-1| = 0$
$\ln|-2| = \ln 2$

(b) $$\int_{-1}^{0} f(x)\mathrm{d}x = \int_{-1}^{0}\left(-\frac{1}{x+2} + \frac{1}{x-1} + \frac{2}{(x-1)^2}\right)\mathrm{d}x$$

$$= \Big[-\ln|x+2| + \ln|x-1| - 2(x-1)^{-1}\Big]_{-1}^{0}$$

$$= -\ln 2 + \ln|-1| + 2 - (-\ln 1 + \ln|-2| + 1)$$

$$= -\ln 2 + 2 - \ln 2 - 1$$

$$= 1 - 2\ln 2$$

2

Expand $(1-2x)^{-3}$ in ascending powers of x as far as the term in x^3 and state the set of values of x for which the expansion is valid.

Take care with the negatives.

Remember to include the restriction on x, even when it is not requested specifically in the question.

$$(1-2x)^{-3} = 1 + (-3)(-2x) + \frac{(-3)(-4)}{2!}(-2x)^2 + \frac{(-3)(-4)(-5)}{3!}(-2x)^3 + \ldots$$

$$= 1 + 6x + 24x^2 + 80x^3 + \ldots \quad \text{provided } |2x| < 1, \text{ i.e. } |x| < 0.5$$

Sample questions and model answers (continued)

3

Use the substitution $x = \tan u$ to show that $\int_0^1 \frac{1}{(1+x^2)^2}\,dx = \int_0^{\frac{\pi}{4}} \cos^2 u\,du.$

Hence show that $\int_0^1 \frac{1}{(1+x^2)^2}\,dx = \frac{\pi}{8} + \frac{1}{4}.$

$$\int_0^1 \frac{1}{(1+x^2)^2}\,dx = \int_{x=0}^{x=1} \frac{1}{(1+x^2)^2}\frac{dx}{du}\,du$$

Let $x = \tan u$

Show this essential side working

Remember to change the x limits to u limits.

$$= \int_0^{\frac{\pi}{4}} \frac{1}{(\sec^2 u)^2}\sec^2 u\,du$$

$$\Rightarrow \frac{dx}{du} = \sec^2 u$$

$$= \int_0^{\frac{\pi}{4}} \frac{1}{\sec^2 u}\,du$$

$$1 + x^2 = 1 + \tan^2 u = \sec^2 u$$

To integrate even powers of $\sin x$ or $\cos x$, change to double angles using $\cos 2x$ formula.

$$= \int_0^{\frac{\pi}{4}} \cos^2 u\,du$$

Limits:

When $x = 0$, $\tan u = 0 \Rightarrow u = 0$

When $x = 1$, $\tan u = 1 \Rightarrow u = \frac{\pi}{4}$

$$\int_0^{\frac{\pi}{4}} \cos^2 u\,du = \frac{1}{2}\int_0^{\frac{\pi}{4}} (1 + \cos 2u)\,du$$

$$\cos 2u = 2\cos^2 u - 1$$

$$= \frac{1}{2}\left[u + \frac{1}{2}\sin 2u\right]_0^{\frac{\pi}{4}}$$

$$\Rightarrow \cos^2 u = \tfrac{1}{2}(1 + \cos 2u)$$

$$= \frac{1}{2}\left(\frac{\pi}{4} + \frac{1}{2} - 0\right)$$

$$= \frac{\pi}{8} + \frac{1}{4}$$

4

(a) Find the centre, C, and radius, r, of the circle $4x^2 + 4y^2 = 39 + 12x - 16y$

(b) Find, by calculation, whether the point $P(4, 1)$ lies inside, on or outside the circle.

Divide through by 4 so that the coefficients of x^2 and y^2 are 1. Then complete the square for the x terms and the y terms.

(a)
$$4x^2 + 4y^2 = 39 + 12x - 16y$$
$$\Rightarrow 4x^2 + 4y^2 - 12x + 16y = 39$$
$$\Rightarrow x^2 - 3x + y^2 + 4y = \tfrac{39}{4}$$
$$\Rightarrow (x - \tfrac{3}{2})^2 - \tfrac{9}{4} + (y + 2)^2 - 4 = \tfrac{39}{4}$$
$$\Rightarrow (x - \tfrac{3}{2})^2 + (y + 2)^2 = 16$$

The circle has centre $C(\tfrac{3}{2}, -2)$ with $r = 4$.

(b) $CP^2 = (4 - \tfrac{3}{2})^2 + (1 - (-2))^2$

$= 15.25$

Since $CP^2 < r^2$, P lies inside the circle.

Sample questions and model answers (continued)

5

The equation of a straight line l is

$\mathbf{r} = 4\mathbf{i} - 2\mathbf{k} + t(3\mathbf{i} + \mathbf{k})$ where t is a scalar parameter.

Points A, B and C lie on the line l. Referred to the origin O, they have position vectors **a**, **b** and **c** respectively.

(a) Find **a** and **b** and angle AOB, given that distance $OA = OB = 10$ units.

(b) Find **c**, given that OC is perpendicular to l.

(a) $\underline{a} = (4 + 3t)\underline{i} + (-2 + t)\underline{k}$ for some value t. (1)

Distance $OA = |\mathbf{a}|$

Since $|\underline{a}| - 10$,

$$\sqrt{(4 + 3t)^2 + (-2 + t)^2} = 10$$

$$16 + 24t + 9t^2 + 4 - 4t + t^2 = 100$$

$$10t^2 + 20t - 80 = 0$$

$$t^2 + 2t - 8 = 0$$

$$(t - 2)(t + 4) = 0$$

$\Rightarrow t = 2$ or $t = -4$

This gives the position vectors of **a** and **b**

Substituting for t in (1) gives

$\underline{a} = 10\underline{i}$ and $\underline{b} = -8\underline{i} - 6\underline{k}$

Let angle AOB be θ.

$|\underline{a}| = |\underline{b}| = 10$ and $\underline{a} . \underline{b} = (10\underline{i}) . (-8\underline{i} - 6\underline{k}) = -80$

$$\underline{a} . \underline{b} = |\underline{a}|\,|\underline{b}| \cos\theta$$

$$\Rightarrow \quad \cos\theta = \frac{-80}{100} = -0.8$$

$$\Rightarrow \quad \theta = 143.1° \text{ (1 d.p.)}$$

(b) $\underline{c} = (4 + 3t)\underline{i} + (-2 + t)\underline{k}$ for some value t. (2)

$3\mathbf{i} + \mathbf{k}$ is the direction vector of l and the scalar product of perpendicular vectors is zero.

Since $\underline{c}$ *is perpendicular to* l,

$((4 + 3t)\underline{i} + (-2 + t)\underline{k}).(3\underline{i} + \underline{k}) = 0$

$\Rightarrow \quad 12 + 9t - 2 + t = 0$

$\Rightarrow \quad t = -1$

Substituting for t in (2) gives $\underline{c} = \underline{i} - 3\underline{k}$.

Practice examination questions

1 It is given that $f(x) = \dfrac{1}{(1+x)^2} + \sqrt{4+x}$.

Show that if x^3 and higher powers are ignored, $f(x) \approx a + bx + cx^2$ and find the values of a, b and c.

State the values of x for which the expansion is valid.

2 (a) Find $\dfrac{dy}{dx}$ when $y = \cos^2 x \sin x$, expressing your answer in terms of $\cos x$.

(b) Using the substitution $u = \cos x$, or otherwise, find $\int \cos^2 x \sin x dx$.

3 A curve has equation $y = \dfrac{x^2 - 4}{x + 1}$.

The equation of the normal to the curve at (2, 0) is $ax + by + c = 0$, where a, b and c are integers.

Find the values of a, b, and c.

4

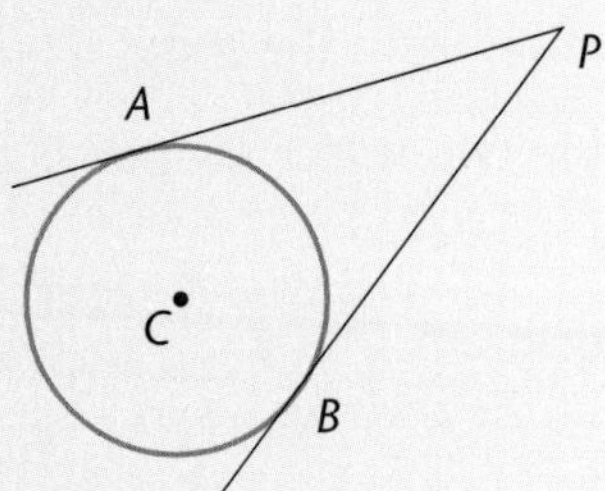

(a) Find, in Cartesian form, the equation of the circle with centre $C(1, 3)$ and radius 2.

(b) The tangents to the circle from $P(5, 8)$ meet the circle at A and B. Find the lengths of AP and BP.

5 Expressing y in terms of x, solve the differential equation

$$x\frac{dy}{dx} = y + yx,$$

given that $y = 4$ when $x = 2$.

6 The equation $x^3 - 2x^2 + x - 4 = 0$ has just one root, α.

(a) Show that $2 < \alpha < 3$.

(b) Taking $x = 2$ as a first approximation, use the Newton-Raphson method to find the value of α, correct to 1 decimal place.

7 (a) Use integration by parts to evaluate $\int_0^1 xe^{-2x}\, dx$.

Practice examination questions (continued)

(b) The diagram shows the graph of $y = xe^{-x}$.

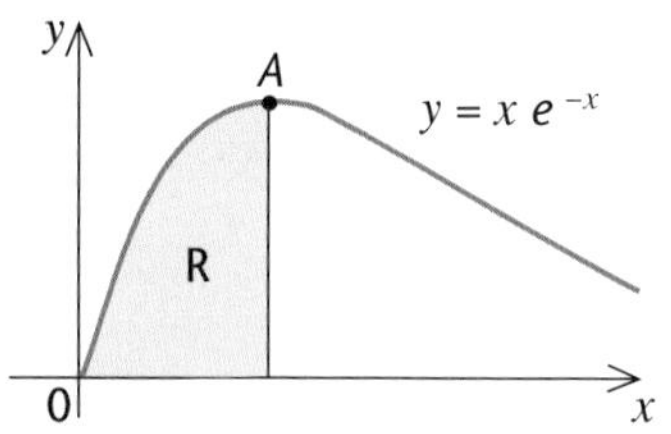

The curve has a maximum point at A.

The shaded region R is the area bounded by the curve, the x-axis and the vertical line through A.

(i) Show that the coordinates of A are $(1, e^{-1})$.
(ii) The region R is rotated completely about the x-axis.
Calculate the volume of the solid generated.

8 In a biology experiment, the growth of a population is being investigated.

A simple model is set up in which the rate of increase of the population at time t is proportional to the number, P, in the population at that time.

It is known that when $t = 0$, $P = 500$ and when $t = 10$, $P = 1000$.

Obtain the relationship between P and t and estimate the size of the population, according to the model, when $t = 20$.

9 (a) Find the equation of the normal to the curve $x^3 + xy + y^2 = 7$ at the point (1, 2).

(b) Find the coordinates of the two stationary points on the curve $x = t^2$, $y = t + \dfrac{1}{t}$.

10 The region R, shaded in the diagram, is bounded by the curve $y = \sin 2x$ and the x-axis.

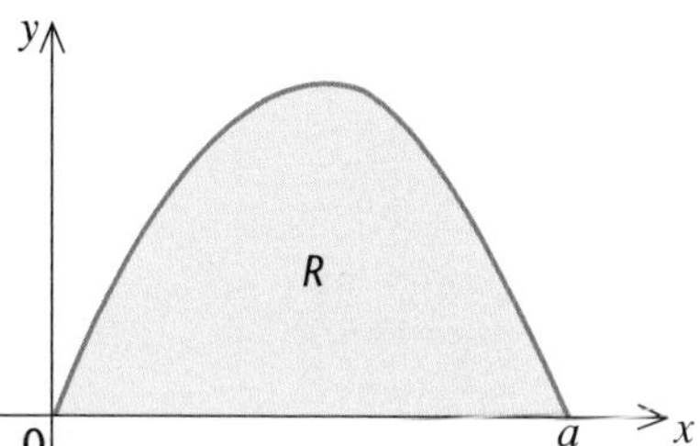

(a) Find the value of a.
(b) Show that the area of R is 1 square unit.
(c) Find the volume of the solid formed when R is rotated completely about the x-axis.

11 (a) Find the exact value of the gradient of the curve $y = \ln(\sec x + \tan x)$ at the point where $x = \dfrac{\pi}{4}$.

(b) Evaluate $\displaystyle\int_0^{\frac{1}{4}\pi} \tan^2 x\,dx$.

12 (a) Find a vector equation of the line AB through $A(2, 2, 0)$ and $B(2, 3, 1)$.

(b) Show that the line AB and the line $\mathbf{r} = 2\mathbf{j} + \mathbf{k} + \lambda(\mathbf{i} + \mathbf{k})$, where λ is a scalar parameter, have no common point.

(c) Find the acute angle between the two lines.

Chapter 2

Statistics 2/3

The following topics are covered in this chapter:

- Continuous random variables
- Statistical approximations
- Estimation and sampling
- Hypothesis tests 1
- Hypothesis tests 2
- χ^2 tests

2.1 Continuous random variables

After studying this section you should be able to:

LEARNING SUMMARY

- find probabilities and the mean and variance of a continuous variable with probability density function $f(x)$
- find the cumulative function $F(x)$ and the median and quartiles
- find the probability density function from the cumulative distribution function
- use the rules relating to linear combinations of variables, especially normal variables
- understand the continuous uniform distribution

The probability density function, $f(x)$

AQA	S4
EDEXCEL	S2
OCR	S2, S3
WJEC	S2, MS

The continuous random variable X is defined by its **probability density function** $f(x)$, together with the values for which it is valid.

$$P(c \leqslant x \leqslant d) = \int_c^d f(x)\mathrm{d}x$$

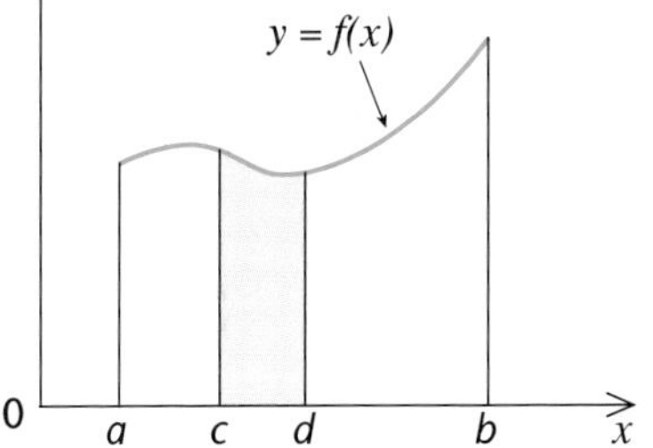

The total probability is 1.

If $f(x)$ is valid for $a \leqslant x \leqslant b$, then $\int_a^b f(x)\mathrm{d}x = 1$.

Key points from AS

- **Continuous random variables**
  Revise AS page 102

Probabilities are given by areas under the curve.

The functions used most often are X and X^2 as E(X) is the mean and E(X^2) is used to find the variance.

The **expectation** of $g(x)$ is $\mathrm{E}(g(x)) = \int_{\text{all } x} g(x)f(x)\mathrm{d}x$, for any function $g(x)$.

The mean is the **expectation** of X, written E(X). It is often denoted by μ.

If $f(x)$ is given in stages, then the integration is carried out in stages also.

KEY POINT

Mean: $\mathrm{E}(X) = \mu = \int_{\text{all } x} xf(x)\mathrm{d}x$

Variance: $\mathrm{Var}(X) = \sigma^2 = \mathrm{E}(X^2) - \mu^2$, where $\mathrm{E}(X^2) = \int_{\text{all } x} x^2 f(x)\mathrm{d}x$

Example

The continuous random variable X has probability density function $f(x)$ given by $f(x) = \frac{1}{4}x$ for $1 \leqslant x \leqslant 3$. Find the mean μ, E(X^2) and the standard deviation σ.

$$\mu = \mathrm{E}(X) = \int_{\text{all } x} xf(x)\mathrm{d}x = \int_1^3 \tfrac{1}{4}x^2\mathrm{d}x = \left[\frac{x^3}{12}\right]_1^3 = 2\tfrac{1}{6}$$

$$\mathrm{E}(X^2) = \int_{\text{all } x} x^2 f(x)\mathrm{d}x = \int_1^3 \tfrac{1}{4}x^3\mathrm{d}x = \left[\frac{x^4}{16}\right]_1^3 = 5$$

$\sigma = \sqrt{\text{variance}}$

$$\mathrm{Var}(X) = 5 - (2\tfrac{1}{6})^2 = \tfrac{11}{36} \Rightarrow \sigma = \sqrt{\tfrac{11}{36}} = 0.553 \text{ (3 d.p.)}$$

The cumulative distribution function, $F(x)$

EDEXCEL	S2
OCR	S2, S3
WJEC	S1

The **cumulative distribution function**, $F(x)$, gives the probability that X is less than a particular value, i.e. $P(X < x)$. It is found from the area under the probability curve:

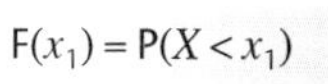

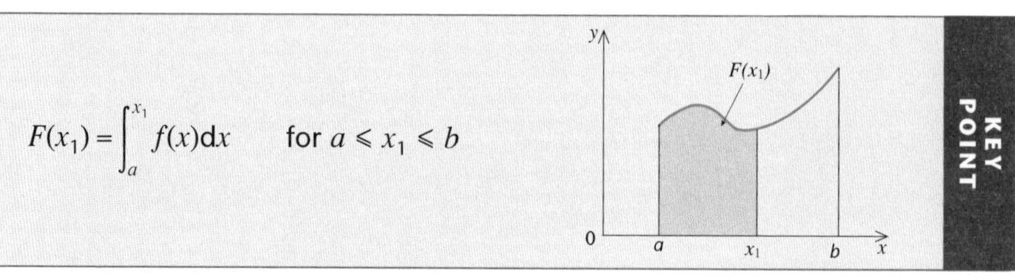

$$F(x_1) = \int_a^{x_1} f(x)\mathrm{d}x \quad \text{for } a \leqslant x_1 \leqslant b$$

Sometimes the median is denoted by q_2.

The **median**, m, is the value 50% of the way through the distribution. It divides the total area in half, so $F(m) = 0.5$.

The **lower quartile** q_1 is the value 25% of the way, so $F(q_1) = 0.25$.
The **upper quartile** q_3 is the value 75% of the way, so $F(q_3) = 0.75$.

The **interquartile range** = upper quartile – lower quartile.

Example

$f(x) = \frac{1}{4}x$ for $1 \leqslant x \leqslant 3$. Find the median, m, and the interquartile range.

x_1 is used here to denote any value of x between 1 and 3. Sometimes a different letter is used, such as t.

For $1 \leqslant x_1 \leqslant 3$, $F(x_1) = \int_1^{x} \frac{1}{4}x\mathrm{d}x = \left[\frac{x^2}{8}\right]_1^{x_1} = \frac{1}{8}(x_1^2 - 1)$

$F(m) = 0.5 \Rightarrow \frac{1}{8}(m^2 - 1) = 0.5 \Rightarrow m^2 = 5,\ m = \sqrt{5}\ (= 2.236\ ...)$

$F(q_1) = 0.25 \Rightarrow \frac{1}{8}(q_1^2 - 1) = 0.25 \Rightarrow q_1^2 = 3,\ q_1 = \sqrt{3}\ (= 1.732\ ...)$

$F(q_3) = 0.75 \Rightarrow \frac{1}{8}(q_3^2 - 1) = 0.75 \Rightarrow q_3^2 = 7,\ q_3 = \sqrt{7}\ (= 2.645\ ...)$

Interquartile range $= q_3 - q_1 = \sqrt{7} - \sqrt{3} = 0.91$ (2 d.p.)

Differentiating the cumulative function gives the probability function.

The probability density function can be obtained from the cumulative distribution function:

KEY POINT

$$f(x) = \frac{\mathrm{d}}{\mathrm{d}x}F(x).$$

Example

Find $f(x)$ where $F(x) = \begin{cases} 0 & x \leqslant 0 \\ \frac{1}{8}x^3 & 0 \leqslant x \leqslant 2 \\ 1 & x \geqslant 2 \end{cases}$

$f(x) = \frac{\mathrm{d}}{\mathrm{d}x}(\frac{1}{8}x^3) = \frac{3}{8}x^2$ for $0 \leqslant x \leqslant 2$ and $f(x) = 0$ otherwise.

Linear combinations of random variables

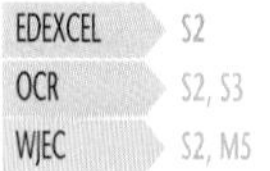

The following results relating to **expectation algebra**, hold for continuous and discrete variables.

In these examples, $b = -4$.

For variable X and constants a and b:	**Examples**
$\mathrm{E}(aX) = a\mathrm{E}(X)$	$\mathrm{E}(3X) = 3\mathrm{E}(X)$
$\mathrm{E}(aX + b) = a\mathrm{E}(X) + b$	$\mathrm{E}(5X - 4) = 5\mathrm{E}(X) - 4$
$\mathrm{Var}(aX) = a^2\,\mathrm{Var}(X)$	$\mathrm{Var}(3X) = 9\,\mathrm{Var}(X)$
$\mathrm{Var}(aX + b) = a^2\,\mathrm{Var}(X)$	$\mathrm{Var}(5X - 4) = 25\,\mathrm{Var}(X)$

Remember the + sign in the variance result for $\text{Var}(aX - bY)$.

For variables X and Y:
$E(aX + bY) = aE(X) + bE(Y)$
$E(aX - bY) = aE(X) - bE(Y)$
If X and Y are independent:
$\text{Var}(aX + bY) = a^2\,\text{Var}(X) + b^2\,\text{Var}(Y)$
$\text{Var}(aX - bY) = a^2\,\text{Var}(X) + b^2\,\text{Var}(Y)$

Examples
$E(2X + 3Y) = 2E(X) + 3E(Y)$
$E(2X - 3Y) = 2E(X) - 3E(Y)$

$\text{Var}(2X + 3Y) = 4\,\text{Var}(X) + 9\,\text{Var}(Y)$
$\text{Var}(2X - 3Y) = 4\,\text{Var}(X) + 9\,\text{Var}(Y)$

Learn this very important result.

If **X and Y are normally distributed**, then sums, differences and multiples of these normal variables are **also normally distributed.**

In general, if $X \sim N(\mu_1, \sigma_1^2)$, $Y \sim N(\mu_2, \sigma_2^2)$ then

$X + Y \sim N(\mu_1 + \mu_2, \sigma_1^2 + \sigma_2^2)$ (sum)

$X - Y \sim N(\mu_1 - \mu_2, \sigma_1^2 + \sigma_2^2)$ (difference)

$aX \sim N(a\mu_1, a^2\sigma_1^2)$ (multiple)

Example

$X \sim N(10, 4)$ and $Y \sim N(8, 3)$. Find the distribution of $2X - 3Y$.

$E(2X - 3Y) = 2E(X) - 3E(Y) = 20 - 24 = -4$
$\text{Var}(2X - 3Y) = 4\,\text{Var}(X) + 9\,\text{Var}(Y) = 16 + 27 = 43$
So $2X - 3Y \sim N(-4, 43)$

The uniform distribution

AQA S4
EDEXCEL S2
WJEC S2, M5

KEY POINT

If $f(x)$ is distributed **uniformly** in the interval $a \leqslant x \leqslant b$, then $f(x) = \dfrac{1}{b-a}$.

f(x)
$\dfrac{1}{b-a}$
0
a
b
x

By symmetry, the mean is mid-way between a and b, i.e. $E(X) = \frac{1}{2}(a + b)$.
It can be shown that $\text{Var}(X) = \frac{1}{12}(b - a)^2$.

Progress check

1 $f(x) = 1 - 0.5x$, $0 \leqslant x \leqslant 2$.
  (a) Find $E(X)$, $E(X^2)$ and $\text{Var}(X)$.
  (b) Find the cumulative function $F(x)$, the median and the interquartile range.

2 $f(x)$ is distributed uniformly in the interval $1 \leqslant x \leqslant 6$. Find:
  (a) $P(1.8 < X < 3.2)$ (b) $E(X)$ (c) the standard deviation of X.

3 $X \sim N(30, 4)$ and $Y \sim N(20, 5)$. Find:
  (a) $P(X + Y > 56)$ (b) $P(X - Y < 7)$ (c) $P(3X - 4Y > -10)$

4 The masses of apples from a particular tree are normally distributed with mean 135 g and standard deviation 5 g. Find the probability that 7 apples from the tree weigh more than 1 kg.

1 (a) $\frac{2}{3}, \frac{2}{3}, \frac{2}{9}$ (b) $F(x) = 0$, $x < 0$, $F(x) = x - 0.25x^2$, $0 \leqslant x \leqslant 2$, $F(x) = 1$, $x > 2$; median = 0.586 (3 d.p.), IQR = 0.732 (3 d.p.)
2 (a) 0.28 (b) 3.5 (c) 1.44 (2 d.p.)
3 (a) 0.0228 (b) 0.1587 (c) 0.9683
4 0.468

2.2 Statistical approximations

After studying this section you should be able to:

- use the Poisson approximation to the binomial distribution
- use the normal approximation to the binomial distribution
- use the normal approximation to the Poisson distribution

LEARNING SUMMARY

The Poisson approximation to the binomial distribution

EDEXCEL	S2
OCR	S2
WJEC	S1

The binomial distribution $X \sim \text{B}(n,p)$ is used to model the number of successes in n independent trials when the probability of success, p, is constant.

The mean of the binomial distribution is np and the variance is npq (where $q = 1 - p$).

With these conditions, $np \approx npq$. This fits in with the Poisson distribution where the mean and the variance are equal.

> When n is large ($n > 50$, say) and p is small ($p < 0.1$, say), $X \sim \text{B}(n,p)$ can be **approximated** by a Poisson distribution with the same mean, where $X \sim \text{Po}(np)$ approximately.
>
> KEY POINT

Key points from AS

- **Binomial distribution** *Revise AS pages 100*
- **Poisson distribution** *Revise AS page 101*

Example

The proportion of defective items produced by a machine is 2.5%. Find the probability of obtaining fewer than 3 defective items in a random sample of 100 items.

If X is the number of defective items in 100, then $X \sim \text{B}(100, 0.025)$.
The mean $np = 100 \times 0.025 = 2.5$.
Since n is large and p is small, use the Poisson approximation, $X \sim \text{Po}(2.5)$.

$\text{P}(X = x) = \frac{\lambda^x}{x!}e^{-\lambda}$

$$\text{P}(X < 3) = \text{P}(X = 0) + \text{P}(X = 1) + \text{P}(X = 2)$$

$$= e^{-2.5} + 2.5e^{-2.5} + \frac{2.5^2}{2!}e^{-2.5} = e^{-2.5}\left(1 + 2.5 + \frac{2.5^2}{2!}\right) = 0.54 \text{ (2 s.f.)}$$

It is often quicker to use cumulative probability tables, so make sure that you know how to use them.

If you have access to cumulative Poisson tables giving $\text{P}(X \leqslant r)$ for various values of r, then these tables can be used with $\lambda = 2.5$ and
$\text{P}(X < 3) = \text{P}(X \leqslant 2) = 0.5438 = 0.54$ (2 s.f.).

The normal approximation to the binomial distribution

EDEXCEL	S2
OCR	S2
WJEC	S2, MS
NICCEA	S3

> When n and p are such that $np > 5$ and $nq > 5$, $X \sim \text{B}(n,p)$ can be **approximated** by a normal distribution with the same mean and variance, where $X \sim \text{N}(np, npq)$ approximately.
>
> KEY POINT

Key points from AS

- **The normal distribution** *Revise AS pages 102–103*

The larger the value of n and the closer that p is to 0.5, the better the approximation.

The normal distribution is continuous, whereas the binomial is discrete, so a **continuity correction** must be used.

Example

The probability that a person supports a particular organisation is 0.3. In a random sample of 50 people, find the probability that at least 20 support the organisation.

Always define the variable, giving its distribution if known.

If X is the number of people in 50 who support the organisation, then $X \sim \text{B}(50, 0.3)$.

Mean $np = 50 \times 0.3 = 15$, variance $npq = 50 \times 0.3 \times 0.7 = 10.5$.
Conditions for normal approximation: $np = 15 > 5$, $nq = 50 \times 0.7 = 35 > 5$

$X \sim \mathrm{N}(\mu, \sigma^2)$ with $\mu = 15$ and $\sigma^2 = 10.5 \Rightarrow \sigma = \sqrt{10.5}$

$\therefore \quad X \sim \mathrm{N}(15, 10.5)$ approximately.

Think of $X = 20$ as going from 19.5 to 20.5. Since 'at least 20' includes $X = 20$, find $\mathrm{P}(X > 19.5)$ in the normal distribution.

Applying the continuity correction,
$\mathrm{P}(X \geqslant 20)$ becomes $\mathrm{P}(X > 19.5)$.

Z: 0 1.389

$$\begin{aligned}\mathrm{P}(X > 19.5) &= \mathrm{P}\left(Z > \frac{19.5 - 15}{\sqrt{10.5}}\right)\\ &= \mathrm{P}(Z > 1.389)\\ &= 1 - \Phi(1.389) = 1 - 0.9176 = 0.082 \text{ (2 s.f.)}\end{aligned}$$

Standardise $Z = \dfrac{X - \mu}{\sigma}$ and then use standard normal tables.

The normal approximation to the Poisson distribution

EDEXCEL S2
OCR S2
WJEC S2, M5
NICCEA S3

The Poisson distribution, $X \sim \mathrm{Po}(\lambda)$, is used to model the number of occurrences

> **KEY POINT**
> When λ is large ($\lambda > 15$ say), $X \sim \mathrm{Po}(\lambda)$ can be approximated by a normal distribution with the same mean and variance, where $X \sim \mathrm{N}(\lambda, \lambda)$ approximately.

of an event, when events occur randomly. The mean is λ and the variance is λ. Since the Poisson distribution is discrete, a continuity correction must be used.

Example

If $X \sim \mathrm{Po}(20)$, find $\mathrm{P}(15 < X < 22)$.

Since n is large, $X \sim \mathrm{N}(20, 20)$ approximately.

You no not want to include 15 or 22 so go from 15.5 to 21.5.

$\mathrm{P}(15 < X < 22)$ becomes $\mathrm{P}(15.5 < X < 21.5)$.

Apply the continuity correction.

Standardise the variables and then use normal tables.

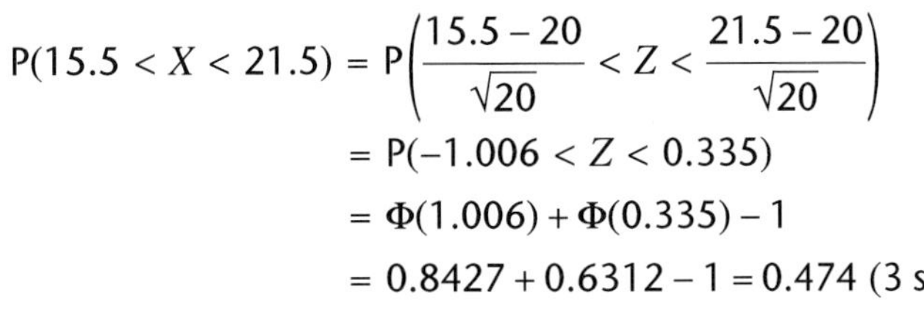

$$\begin{aligned}\mathrm{P}(15.5 < X < 21.5) &= \mathrm{P}\left(\frac{15.5 - 20}{\sqrt{20}} < Z < \frac{21.5 - 20}{\sqrt{20}}\right)\\ &= \mathrm{P}(-1.006 < Z < 0.335)\\ &= \Phi(1.006) + \Phi(0.335) - 1\\ &= 0.8427 + 0.6312 - 1 = 0.474 \text{ (3 s.f.)}\end{aligned}$$

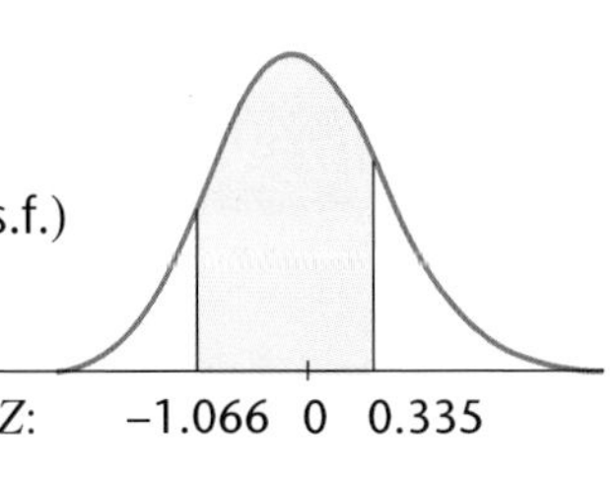

Progress check

1 $X \sim \mathrm{B}(80, 0.06)$. Find, to 3 decimal places, $\mathrm{P}(X = 4)$
   (a) by using the binomial distribution,
   (b) by using a suitable approximation.

2 A newspaper reports that 62% of adults have an e-mail address. A random sample of 40 adults was selected. Using a suitable approximation, find the probability that at least 30 had an e-mail address.

3 Telephone calls reach a switchboard independently and at random at a rate of 30 per hour. Find the probability that in a randomly selected period of one hour there are fewer than 20 calls.

1 (a) 0.186 (b) 0.182 **2** 0.063 (2 s.f.) **3** 0.028 (2 s.f.)

2.3 Estimation and sampling

After studying this section you should be able to:

- *distinguish between different sampling techniques*
- *find unbiased estimates for population parameters from a sample*
- *understand the theory relating to sampling distributions*
- *use the central limit theorem*
- *find confidence intervals*

LEARNING SUMMARY

Sampling techniques

EDEXCEL S2, S3
OCR S2
NICCEA S2/3

When you want information about a **population**, you could carry out a **census** of every member. The advantage is that you would have accurate and complete information. Disadvantages include cost and time requirements. Taking a census could also destroy the population. For example, if you were investigating the length of life of a calculator battery, carrying out a census would destroy the population.

More usually, a **sample** is taken. The sample should be representative of the whole population, so **bias** in the choice of sample members must be avoided.

Assigning a number to each member of a population and then drawing the numbers out of a hat is one method of obtaining a **simple random sample**. Another is to use **random number tables**, in which each digit has an equal chance of occurring.

If there is a periodic fault, then systematic sampling may fail to register it or may give it undue weight.

Random sampling from a very large population can be very laborious. It may be more convenient to carry out **systematic sampling**. This involves selecting every kth member of the population, for example, every 10th item from a particular machine on a production line.

The stratified sample can be constructed in proportion to the number of members in each stratum.

The method of **stratified sampling** is often used when the population is split into distinguishable layers or strata, such as students in each faculty in a college.

Unbiased estimates

AQA S4
EDEXCEL S3
OCR S2
WJEC S3
NICCEA S2

When a **population parameter**, such as the mean, variance or proportion, is **unknown**, then it is sensible to estimate it from a sample.

An **unbiased estimate** is one which, on the average, gives the true value, i.e. E(estimate) = true value of parameter.

The best unbiased estimate is the estimate with the smallest variance.

Best unbiased estimates:

For mean μ	$\hat{\mu} = \bar{x} = \frac{\Sigma x}{n}$	$\bar{x}$ is sample mean
For proportion p	$\hat{p} = p_s$	p_s is sample proportion
For variance σ^2	$\hat{\sigma}^2 = \frac{n}{n-1}s^2$	s^2 is the sample variance.

Key points from AS

- **Mean, variance and standard deviation** *Revise AS pages 94–95*

Make sure that you are familiar with the format given in your examination booklet and practise using it.

Alternative formats: $\hat{\sigma}^2 = \frac{\Sigma(x-\bar{x})^2}{n-1}$, $\hat{\sigma}^2 = \frac{1}{n-1}\left(\Sigma x^2 - \frac{(\Sigma x)^2}{n}\right)$.

If you are given only summary data such as Σx or Σx^2, substitute into the appropriate formula.

A calculator in statistical mode can be used to obtain the value of $\hat{\sigma}$ directly from raw data. Look for the key marked $\sigma_{x_{n-1}}$.

Sampling distributions

AQA	S1, S4, S5
EDEXCEL	S3
OCR	S2
WJEC	S2, S3
NICCEA	S2

The most important sampling distribution is the distribution of the sample mean and you will need to be able to use it to find probabilities.

The other sampling distributions are included for reference, as they provide the basis for confidence intervals and the test statistic in hypothesis testing. Questions based directly on them, however, are unlikely to be included in the examination.

The sampling distribution of the mean

If all possible samples of size n are taken from $X \sim N(\mu, \sigma^2)$ and their sample means calculated, then these means form the sampling distribution of means, $\bar{X}$ which is also normally distributed.

This important result can be derived using expectation algebra (p. 53).

KEY POINT

If $X \sim N(\mu, \sigma^2)$, then $\bar{X} \sim N\left(\mu, \frac{\sigma^2}{n}\right)$.

The mean of $\bar{X}$ is the same as the mean of X.

The variance of $\bar{X}$ is much smaller than the variance of X.

Standard deviation = $\sqrt{\text{variance}}$

The diagram shows the curves for X and for $\bar{X}$.

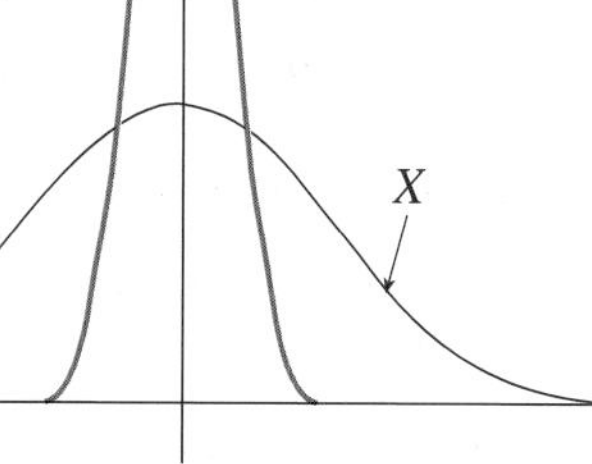

They are both symmetrical about μ.
The curve for $\bar{X}$ is much more squashed in, confirming the smaller standard deviation.

The standard deviation of the sampling distribution, $\sigma/\sqrt{n}$, is known as the **standard error of the mean.**

The following result is extremely useful and should be learnt.

KEY POINT

If the distribution of X is not normal, then, by the central limit theorem, $\bar{X} \sim N\left(\mu, \frac{\sigma^2}{n}\right)$ approximately, provided n is large.

$\hat{\sigma}^2$ is the unbiased estimate of σ^2.

If you do not know σ^2, the variance of X, then it can be estimated by $\hat{\sigma}^2$ and provided n is large, $\bar{X} \sim N\left(\mu, \frac{\hat{\sigma}^2}{n}\right)$ approximately.

Example

The heights of men in a particular area are normally distributed with mean 176 cm and standard deviation 7 cm. A random sample of 50 men is taken. Find the probability that the mean height of the men in the sample is less than 175 cm.

If X is the height, in centimetres, of a man from this area, then $X \sim N(176, 7^2)$.

For random samples of size 50, $\bar{X} \sim N\left(176, \frac{7^2}{50}\right)$, i.e. $\bar{X} \sim N(176, 0.98)$.

In this example, since the population is normal, any size sample, large or small, could have been taken.

$$\begin{aligned} P(\bar{X} < 175) &= P\left(Z < \frac{175 - 176}{\sqrt{0.98}}\right) \\ &= P(Z < -1.010\ldots) \\ &= 1 - \Phi(1.010) \\ &= 1 - 0.8438 \\ &= 0.156 \text{ (3 d.p.)} \end{aligned}$$

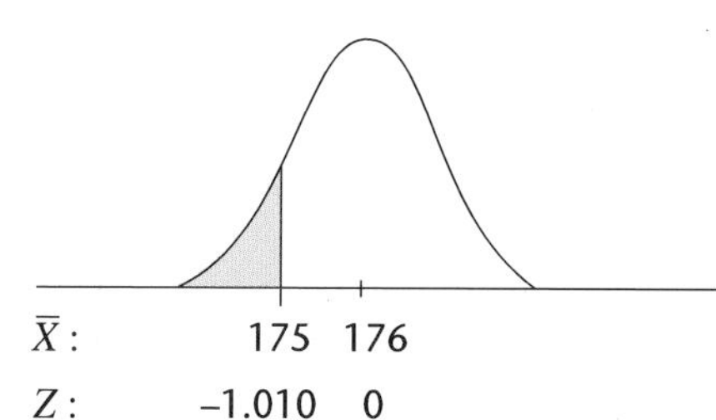

X and Y must be independent.

If X and Y are not normally distributed, then use the central limit theorem if samples are large.

The sampling distribution of the difference between means

Taking random samples of size n_1 from $X \sim N(\mu_1, \sigma_1^2)$ and size n_2 from $Y \sim N(\mu_2, \sigma_2^2)$, the sampling distribution of the difference between means is also normally distributed, where $\bar{X} - \bar{Y} \sim N\left(\mu_1 - \mu_2, \frac{\sigma_1^2}{n_1} + \frac{\sigma_2^2}{n_2}\right)$.

If σ_1^2 and σ_2^2 are unknown, they can be estimated using $\hat{\sigma}_1^2$ and $\hat{\sigma}_2^2$, provided n_1 and n_2 are large.

If X and Y have common variance σ^2, then $\bar{X} - \bar{Y} \sim N\left(\mu_1 - \mu_2, \sigma^2\left(\frac{1}{n_1} + \frac{1}{n_2}\right)\right)$.

If σ^2 is unknown, then it can be estimated using

s_1^2 and s_2^2 are the sample variances.

$$\hat{\sigma}^2 = \frac{n_1 s_1^2 + n_2 s_2^2}{n_1 + n_2 - 2} = \frac{(n_1 - 1)\hat{\sigma}_1^2 + (n_2 - 1)\hat{\sigma}_2^2}{n_1 + n_2 - 2}$$

$\hat{\sigma}_1^2$ and $\hat{\sigma}_2^2$ are the unbiased estimates of σ_1^2 and σ_2^2.

If n_1 and n_2 are small, the t distribution is needed. In this case, X and Y must be normally distributed (see section 2.5).

$X - Y \sim N\left(\mu_1 - \mu_2, \hat{\sigma}^2\left(\frac{1}{n_1} + \frac{1}{n_2}\right)\right)$ approximately, provided n_1 and n_2 are large.

The sampling distribution of proportions

Consider a population with proportion of successes, p.

The normal approximation to the binomial has been used to derive this.

If samples of size n (large) are taken, the sampling distribution of proportions, $P_s \sim N\left(p, \frac{pq}{n}\right)$, approximately.

The standard deviation, $\sqrt{pq/n}$ is known as the **standard error of proportion.**

When using this distribution, a continuity correction of $\pm \frac{1}{2n}$ is needed.

The sampling distribution of the difference between proportions

n_1 and n_2 should be large.

If samples of size n_1 are taken from X (with proportion p_1) and samples of size n_2 taken from Y (with proportion p_2), then the sampling distribution of the difference between proportions, $P_{s_1} - P_{s_2} \sim N\left(p_1 - p_2, \left(\frac{p_1 q_1}{n_1} + \frac{p_2 q_2}{n_2}\right)\right)$.

If both populations have common proportion p, it can be estimated using $\hat{p} = \frac{n_1 P_{s_1} + n_2 P_{s_2}}{n_1 + n_2}$. In this case, $P_{s_1} - P_{s_2} \sim N\left(0, \hat{p}\hat{q}\left(\frac{1}{n_1} + \frac{1}{n_2}\right)\right)$.

Confidence intervals 1

AQA	S4, S5
EDEXCEL	S3
OCR	S3
WJEC	S3
NICCEA	S2, S3

A confidence interval for an unknown population parameter is an interval (a, b) such that there is a specified probability, often 90%, 95% or 99%, that the interval contains the true value.

Confidence interval for the mean μ

This is calculated using the mean, $\bar{x}$, of a random sample of size n.
If the population is normal, the sample can be any size. If it is not normal, then n must be large.

±1.96 are the standardised z-values for the central 95% of a standard normal distribution.

Provided n is large, $\hat{\sigma}^2$ can be used if σ^2 is unknown.

The 95% **confidence limits** for μ are $\bar{x} \pm 1.96\dfrac{\sigma}{\sqrt{n}}$.

The 95% **confidence interval** is $\left(\bar{x} - 1.96\dfrac{\sigma}{\sqrt{n}},\ x + 1.96\dfrac{\sigma}{\sqrt{n}}\right)$.

$\bar{x}$

$\bar{x} - 1.96\dfrac{\sigma}{\sqrt{n}}$ $\qquad$ $\bar{x} + 1.96\dfrac{\sigma}{\sqrt{n}}$

$\longleftarrow\ 2 \times 1.96\dfrac{\sigma}{\sqrt{n}}\ \longrightarrow$

The **width** of the 95% confidence interval is $2 \times 1.96\dfrac{\sigma}{\sqrt{n}}$.

Different levels of confidence

±1.645 are the standardised z-values for the central 90% of a standard normal distribution.

±2.576 are the standardised z-values for the central 99% of a standard normal distribution.

Any level of confidence can be used. The most common are 95%, (described above), 90% and 99%.

90% confidence interval is $\left(\bar{x} - 1.645\dfrac{\sigma}{\sqrt{n}},\ \bar{x} + 1.645\dfrac{\sigma}{\sqrt{n}}\right)$.

99% confidence interval is $\left(\bar{x} - 2.576\dfrac{\sigma}{\sqrt{n}},\ \bar{x} + 2.576\dfrac{\sigma}{\sqrt{n}}\right)$.

Example

X is the mass, in grams, of a bag of flour packed by a particular machine. The mean mass of bags packed by this machine is μ and the standard deviation is σ. A random sample of 100 bags gave the following results:

$$\Sigma x = 50\,140,\ \Sigma x^2 = 25\,166\,580.$$

Find a 95% confidence interval for μ.

$$\bar{x} = \frac{\Sigma x}{n} = \frac{50\,140}{100} = 501.4$$

Since σ^2 is unknown, find $\hat{\sigma}^2$

$$\hat{\sigma}^2 = \frac{1}{n-1}\left(\Sigma x^2 - \frac{(\Sigma x)^2}{n}\right)$$

$$= \frac{1}{99}\left(25\,166\,580 - \frac{50\,140^2}{100}\right) = 266.5, \text{ so } \hat{\sigma} = \sqrt{266.5} = 16.32$$

You can work out the confidence interval straight away, but it is often easier to calculate the limits first.

95% confidence limits are $\bar{x} \pm 1.96\dfrac{\hat{\sigma}}{\sqrt{n}} = 501.4 \pm 1.96\dfrac{16.32}{\sqrt{100}} = 501.4 \pm 3.20$

95% confidence interval $= (501.4 - 3.20,\ 501.4 + 3.20)$
$= (498.2 \text{ g},\ 504.6 \text{ g})$ (1 d.p.)

The distribution of X is not known. Since n is large, the central limit theorem has been used in the underlying theory.

Confidence intervals for the difference between means

With the notation used on page 58, the confidence interval is as shown below when σ_1^2 and σ_2^2 are known. It can be amended for the other situations described on page 58.

95% confidence interval: $\left(\bar{x} - \bar{y} - 1.96\sqrt{\dfrac{\sigma_1^2}{n_1} + \dfrac{\sigma_2^2}{n_2}},\ \bar{x} - \bar{y} + 1.96\sqrt{\dfrac{\sigma_1^2}{n_1} + \dfrac{\sigma_2^2}{n_2}}\right)$

Confidence interval for proportion, p

This is calculated using the proportion, p_s in a random sample (large n).

$$95\% \text{ confidence interval is } \left(p_s - 1.96\sqrt{\frac{p_s q_s}{n}}, p_s + 1.96\sqrt{\frac{p_s q_s}{n}}\right).$$

Example

Find a 95% confidence interval for the population proportion p if the proportion of successes in a random sample of 100 is 0.6

The 95% confidence limits for p are $0.6 \pm 1.96\sqrt{\frac{0.6 \times 0.4}{100}} = 0.6 \pm 0.096.$

so the 95% confidence interval is (0.504, 0.696).

This means that the probability that (0.504, 0.696) includes p is 95%.

Confidence intervals for the difference between proportions

With the notation used on page 58

95% confidence interval is

$$\left(p_{s_1} - p_{s_2} - 1.96\sqrt{\frac{p_1 q_1}{n_1} + \frac{p_2 q_2}{n_2}},\ p_{s_1} - p_{s_2} + 1.96\sqrt{\frac{p_1 q_1}{n_1} + \frac{p_2 q_2}{n_2}}\right)$$

Progress check

1 The random variable X has mean μ and standard deviation σ.
Ten independent observations of X are
4.6, 3.9, 4.2, 5.8, 6.3, 4.2, 7.2, 6.0, 7.1, 5.4.

(a) Find unbiased estimates of μ and σ.
(b) Given that X is normally distributed with variance 2, calculate a 99% confidence interval for μ.

2 A random sample of size 100 is taken from a normal distribution with mean 120 and standard deviation 5.

(a) Find $P(118.75 < \bar{X} < 120.5)$, where $\bar{X}$ is the sample mean.
(b) State, with a reason, whether your answer would be different if X is not normally distributed.

3 In a randomly selected sample of 400 television viewers, 312 said that they watched the final episode of a long running series.

(a) Calculate a 95% confidence interval for p, the proportion of television viewers who watched the final episode.
(b) If 60 such surveys were conducted and 95% confidence intervals constructed, how many of these would you expect to contain p?

4 X and Y are normally distributed with variance 25.
A sample of 10 observations from X has mean 35.2 and a sample of 20 observations from Y has mean 32.6.
Calculate a 90% confidence interval for the difference between means.

1 (a) 5.47, 1.21 (b) (4.32, 6.62)
2 0.8351; no, use central limit theorem since n is large
3 (a) (0.74, 0.82) (b) 57
4 (−0.586, 5.786)

2.4 Hypothesis tests 1

After studying this section you should be able to:

- *understand the language used in hypothesis testing*
- *understand Type I and Type II errors*
- *understand how to obtain critical z-values*
- *perform the z-test for the mean and the difference between means*
- *perform tests for a binomial proportion (large and small sample sizes)*
- *perform a discrete test for a Poisson mean (small and large λ)*
- *perform a z-test for the difference between proportions (large samples)*

LEARNING SUMMARY

The language of hypothesis testing

AQA	S4
EDEXCEL	S2
OCR	S2
WJEC	S2, S3
NICCEA	S2

When you want to know something about a population, you might perform a **hypothesis** or **significance test.**

A **null hypothesis,** H_0 is made about the population.
Then the **alternative hypothesis** H_1, the situation when H_0 is not true, is stated.

Example

When investigating the mean of a normal distribution, you might make the null hypothesis that the mean is 5. This is written H_0: $\mu = 5$.

The alternative hypothesis would be one of the following:

Use $\mu > 5$ if you suspect an *increase.*

Use $\mu \neq 5$ if you suspect a *change.*

- H_1: $\mu > 5$ (one-tailed, upper tail test)
- H_1: $\mu < 5$ (one-tailed, lower tail test)
- H_1: $\mu \neq 5$ (two-tailed test).

Use $\mu < 5$ if you suspect a *decrease.*

The formula for the test statistic depends on the sampling distribution. Examples are shown in the following text.

A **test statistic** is defined and its distribution when H_0 is true is stated.
Its value, based on information from a random sample taken from the population, is found. This is the **test value.**

The hypothesis test involves deciding whether or not the test value could have come from the distribution defined by the null hypothesis.
If the test value is in the main bulk of the distribution (the **acceptance region**) it is *likely* to have come from the distribution. If it is in the 'tail' end (the **rejection** or **critical region**), it is *unlikely* to have come from the distribution.

Probability theory is used to decide the placing of the boundary between 'likely' and 'unlikely'.

The decision rule (**rejection criterion**) is based on the **significance level** of the test, which is used to fix the boundaries for the rejection region. The boundaries are called **critical values.**

The significance level is a given as a percentage.

The levels used most often are 10%, 5% and 1%.

For **example**, for a significance level of 5%, the critical values are such that 5% of the distribution is in the critical (rejection) region,

There are two critical values in a two-tailed test.

i.e. P(X > critical value) = 0.05, for a one-tailed, upper tail test
P(X < critical value) = 0.05, for a one-tailed, lower tail test
P(X > upper critical value) = 0.025, for a two-tailed test
P(X < lower critical value) = 0.025, for a two-tailed test.

A test statistic is said to be significant if it lies in the critical region. If it lies in the acceptance region, then it is not significant.

Depending on the position of the test value, the **decision** is made.
If the test value lies in the critical region, H_0 is rejected in favour of H_1.
If the test value is in the acceptance region, H_0 is not rejected.

The **conclusion** is then stated in relation to the situation being tested.

Type I and Type II errors

AQA S5
OCR S2

You make a Type I error when you reject a true null hypothesis.

You make a Type II error when you accept a false null hypothesis.

In a hypothesis test, either H_0 is rejected or H_0 is not rejected. There are occasions when the decision is incorrect and an error is made.

If H_0 is rejected when it is in fact true, a **Type I** error is made.

The probability of making a Type I error is the same as the significance level of the test. For example, if the significance level is 5%, then P(Type I error) = 0.05.

If H_0 is accepted (i.e. not rejected) when it is in fact false, a **Type II** error is made.

To find the probability of making a Type II error, a specific value for H_1 must be given. Then P(Type II error) = P(H_0 is accepted when H_1 is true).

The **power** of a test = 1 – P(Type II error).

Critical values for z-tests

AQA S4, S5
EDEXCEL S3
OCR S2
WJEC S2
NICCEA S2

Some hypothesis tests involving the normal distribution are known as z-tests. In a z-test, the critical values for the rejection region can be found using the standard normal distribution, Z. The most commonly used values are summarised below.

5%
For one-tailed upper tail, $\Phi(z) = 0.95$
For two tailed, upper tail, $\Phi(z) = 0.975$
Use symmetry for lower tails.

10%
For one-tailed upper tail, $\Phi(z) = 0.90$
For two tailed, upper tail $\Phi(z) = 0.95$

1%
For one-tailed upper tail, $\Phi(z) = 0.99$
For two tailed, upper tail $\Phi(z) = 0.995$

Level	*Type of test*	*Critical value*	*Decision rule*
5%	One-tailed (upper)	1.645	Reject H_0 if $z > 1.645$
	One-tailed (lower)	–1.645	Reject H_0 if $z < -1.645$
	Two-tailed	±1.96	Reject H_0 if $z > 1.96$ or $z < -1.96$
10%	One-tailed (upper)	1.282	Reject H_0 if $z > 1.282$
	One-tailed (lower)	–1.282	Reject H_0 if $z < -1.282$
	Two-tailed	±1.645	Reject H_0 if $z > 1.645$ or $z < -1.645$
1%	One-tailed (upper)	2.326	Reject H_0 if $z > 2.326$
	One-tailed (lower)	–2.326	Reject H_0 if $z < -2.326$
	Two-tailed	±2.576	Reject H_0 if $z > 2.576$ or $z < -2.576$

z-test for the mean

AQA S4, S5
EDEXCEL S3
OCR S2
WJEC S2
NICCEA S2

This is used to test the mean μ of a population when you know the variance σ^2. If the population is **normal**, the samples can be **any size**, but if the population is **not normal**, then **large** samples must be taken.

The distribution considered for the test statistic is the sampling distribution of means (see page 57).

When the population is not normal, the central limit theorem is needed.

KEY POINT

Test statistic $Z = \dfrac{\overline{X} - \mu}{\sigma/\sqrt{n}} \sim \mathrm{N}(0, 1)$.

This formula can also be used if the variance of X is not known, provided n is large. In this case, the unbiased estimate $\hat{\sigma}^2$ is used for σ^2.

Example

A machine fills tins of baked beans such that the mass of a filled tin is normally distributed with mean mass 429 g and the standard deviation is 3 g. The machine breaks down and after being repaired, the mean mass of a random sample of 100 tins is found to be 428.45 g. Is this evidence, at the 5% level, that the mean mass of tins filled by this machine has decreased? Assume that the standard deviation remains unaltered.

Define your variables.

Let X be the mass, in grams, of a filled tin and let the mean after the repair be μ.

State the hypotheses.

H_0: $\mu = 429$; H_1: $\mu < 429$.

State the distribution of the test statistic assuming the null hypothesis is true.

Consider the sampling distribution of means $\bar{X} \sim \mathrm{N}\left(429, \dfrac{3^2}{100}\right)$.

State the type of test and the rejection criterion.

Perform a one-tailed (lower tail) test at the 5% level and reject H_0 if $z < -1.645$.

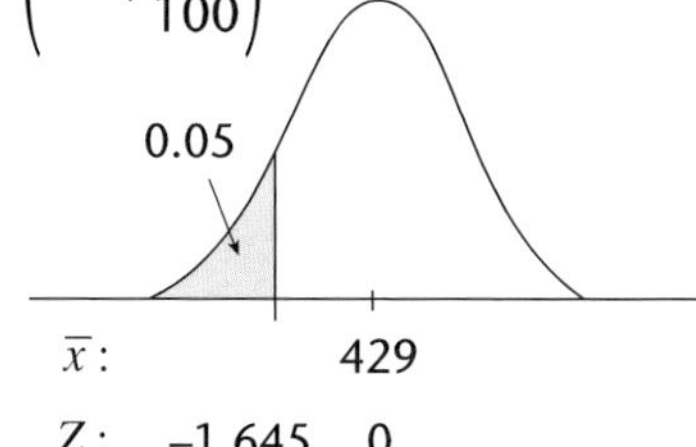

Calculate the value of the test statistic.

$\bar{x} = 428.45$, so

$$z = \frac{\bar{x} - \mu}{\sigma/\sqrt{n}} = \frac{428.45 - 429}{3/\sqrt{100}} = -1.833\ldots$$

Decide whether to reject H_0 or not and relate your conclusion to the question.

Since $z < -1.645$, the test value lies in the critical region and H_0 is rejected.

There is evidence, at the 5% level, that the mean mass of a filled tin has decreased.

Alternatively, a **probability method** can be used.
The rejection rule is to reject H_0 if $\mathrm{P}(\bar{X} < 428.45) < 0.05$.

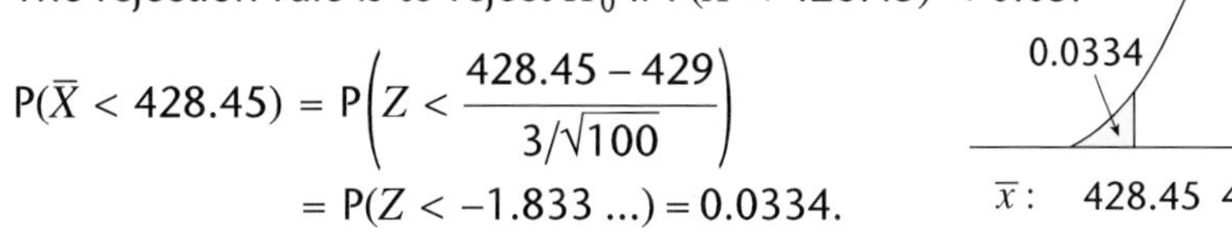

$$\mathrm{P}(\bar{X} < 428.45) = \mathrm{P}\left(Z < \frac{428.45 - 429}{3/\sqrt{100}}\right) = \mathrm{P}(Z < -1.833\ldots) = 0.0334.$$

Since $\mathrm{P}(\bar{X} < 428.45) < 0.05$, H_0 is rejected and the conclusion is as above.

z-test for difference between means

EDEXCEL	S3
OCR	S3
WJEC	S3
NICCEA	S2, S3

This z-test is used to compare the means of two independent normal populations. A sample of size n_1 is taken from $X_1 \sim \mathrm{N}(\mu_1, \sigma_1^2)$ and a sample of size n_2 is taken from $X_2 \sim \mathrm{N}(\mu_2, \sigma_2^2)$.
The sampling distribution for the test statistic is the **difference between means** (page 58).

If σ_1^2 and σ_2^2 are unknown, they can be estimated using $\hat{\sigma}_1^2$ and $\hat{\sigma}_2^2$, provided n_1 and n_2 are large (see p. 58).

KEY POINT

Test statistic $Z = \dfrac{\bar{X}_1 - \bar{X}_2 - (\mu_1 - \mu_2)}{\sqrt{\dfrac{\sigma_1^2}{n_1} + \dfrac{\sigma_2^2}{n_2}}} \sim \mathrm{N}(0, 1)$

If X and Y have common variance σ^2, then

$$Z = \frac{\bar{X}_1 - \bar{X}_2 - (\mu_1 - \mu_2)}{\sigma\sqrt{\dfrac{1}{n_1} + \dfrac{1}{n_2}}} \sim \mathrm{N}(0, 1)$$

If X_1 and X_2 are not normally distributed, then n_1 and n_2 must be large (use central limit theorem).

If σ^2 is unknown, then it can be estimated by $\hat{\sigma}^2$, provided n_1 and n_2 are large (see p. 58)

Tests for a binomial proportion

EDEXCEL S2
OCR S2
WJEC S3

This binomial test is used to test the proportion of successes, p in a binomial population $X \sim B(n, p)$.

Large sample size

See p. 54 Normal approximation to the binomial.

When n is large (such that $np > 5$ and $nq > 5$) a normal approximation to the binomial distribution can be used.

When testing in the upper tail, subtract 0.5, and when testing in the lower tail, add 0.5.

A continuity correction of +0.5 or −0.5 is needed as the binomial distribution is discrete and the normal distribution is continuous.

KEY POINT

Test statistic $Z = \dfrac{(X \pm 0.5) - np}{\sqrt{npq}} \sim N(0, 1)$, provided n is large.

The hypothesis test follows the same pattern as the z-test for the mean.

Example

A coin is tossed 50 times and 30 heads are obtained. Is this evidence, at the 5% level of significance, that the coin is biased in favour of heads?

Define the variable.

Let the probability that the coin shows heads be p.
If X is the number of heads in 50 tosses, then $X \sim B(50, p)$.

State the hypotheses and also the distribution if the null hypothesis is true.

H_0: $p = 0.5$ (the coin is fair), H_1: $p > 0.5$ (the coin is biased in favour of heads).
If the null hypothesis is true, then $X \sim B(50, 0.5)$.

Justify the normal approximation.

But n is large such that $np = 50 \times 0.5 = 25 > 5$ and $nq = 25 > 5$, so $X \sim N(np, npq)$ approximately. Since $npq = 50 \times 0.5 \times 0.5 = 12.5$, $X \sim N(25, 12.5)$.

State the rejection criterion.

At the 5% level, reject H_0 if $z > 1.645$.

29.5 is used because you need to test whether the whole rectangle representing $x = 30$ (from 29.5 to 30.5) is in the upper tail critical region.

The test value is $x = 30$, so using the continuity correction,

$$z = \frac{(x - 0.5) - np}{\sqrt{npq}} = \frac{29.5 - 25}{\sqrt{12.5}} = 1.27 \ldots$$

Since $z < 1.645$, the test value does not lie in the critical region and H_0 is not rejected.

Relate the conclusion to the question.

At the 5% level there is not sufficient evidence to say that the coin is biased in favour of heads.

Alternative method for binomial test, when n is large

Consider the proportion of successes in the sample, p_s.
The distribution for the test statistic is the sampling distribution of proportions.

When working in proportions, the continuity correction is $+\dfrac{1}{2n}$ or $-\dfrac{1}{2n}$.

KEY POINT

Test statistic $Z = \dfrac{\left(p_s \pm \dfrac{1}{2n}\right) - p}{\sqrt{pq/n}} \sim N(0, 1)$, provided n is large.

Example

Using the data for the coin described above:
H_0: $p = 0.5$; H_1: $p > 0.5$.
Using $p = 0.5$, $\sqrt{pq/n} = \sqrt{(0.5 \times 0.5)/50} = \sqrt{0.005}$.

If the continuity correction is used both times (or omitted both times), the z-values agree exactly in the two methods.

The sample proportion $p_s = \frac{30}{50} = 0.6$, so, using the continuity correction,

$$z = \frac{\left(0.6 - \dfrac{1}{2 \times 50}\right) - 0.5}{\sqrt{0.005}} = 1.27 \ldots \text{ as before.}$$

Binomial test when the sample size is small

When n is small, the normal approximation does not apply, so binomial probabilities are used to decide whether the test value is in the critical region or not.

For **example**, at the 5% level, the decision rules are:

- for a one-tailed (upper tail) test, reject H_0 if $P(X \geqslant \text{test value}) < 0.05$
- for a one-tailed (lower tail) test, reject H_0 if $P(X \leqslant \text{test value}) < 0.05$
- for a two-tailed test, reject H_0 if $P(X \leqslant \text{test value}) < 0.025$ or $P(X \geqslant \text{test value}) < 0.025$.

Example

This is a small sample, so the normal approximation cannot be used.

When a die was thrown 10 times, a six occurred 4 times. Is this evidence, at the 10% level of significance, that the die is biased in favour of sixes?

This is a one-tailed (upper tail) test.

Let the probability of obtaining a six be p.
If X is the number of sixes in 10 throws, then $X \sim B(10, p)$.
H_0: $p = \frac{1}{6}$ (the die is fair), H_1: $p > \frac{1}{6}$ (the die is biased in favour of sixes).

If the null hypothesis is true, then $X \sim B(10, \frac{1}{6})$.
Reject H_0 if the test value of $x = 4$ lies in the critical region (the upper tail 10%) i.e. reject H_0 if $P(X \geqslant 4) < 0.1$.

It may be possible to calculate the probability using cumulative binomial tables. Check in your examination booklet.

$$\begin{aligned} P(X \geqslant 4) &= 1 - P(X < 4) \\ &= 1 - ((\tfrac{5}{6})^{10} + {}^{10}C_1(\tfrac{5}{6})^9(\tfrac{1}{6}) + {}^{10}C_2(\tfrac{5}{6})^8(\tfrac{1}{6})^2 + {}^{10}C_3(\tfrac{5}{6})^7(\tfrac{1}{6})^3) \\ &= 1 - 0.930\ldots = 0.070 \text{ (2 s.f.)}. \end{aligned}$$

Since $P(X \geqslant 4) < 0.1$, H_0 is rejected.

There is evidence at the 10% level, that the die is biased in favour of sixes.

z-test for the difference between proportions

To test whether two binomial populations have a common proportion of successes, p, the sampling distribution of proportions is considered.

The null hypothesis is H_0: $p_1 = p_2 = p$.

$\hat{p} = \dfrac{n_1 p_{s_1} + n_2 p_{s_2}}{n_1 + n_2}$ (see p. 58)

KEY POINT

Test statistic $Z = \dfrac{p_{s_1} - p_{s_2} - 0}{\sqrt{\hat{p}\hat{q}\left(\dfrac{1}{n_1} + \dfrac{1}{n_2}\right)}} \sim N(0, 1)$, provided n_1 and n_2 are large.

Test for a Poisson mean

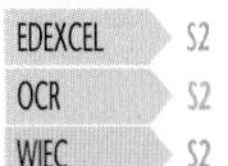

The test statistic is X, the number of occurrences, where $X \sim Po(\lambda)$.

When λ is **small**, the test is similar to the small sample binomial test.

When λ is **large**, $X \sim N(\lambda, \lambda)$ approximately and the test is similar to the large sample binomial test. A continuity correction is needed.

Example

The office manager claims that the average number of telephone calls per minute received by her office switchboard is 4.5. Her supervisor maintains that it is less than 4.5. In a randomly selected minute, 2 calls were received. A hypothesis test is conducted at the 5% level. Whose claim is upheld?

Let X be the number of calls per minute. Assuming that calls occur randomly and independently, $X \sim \text{Po}(\lambda)$.

This is a one-tailed (lower tail) test.

H_0: $\lambda = 4.5$, H_1: $\lambda < 4.5$. If the null hypothesis is true, $X \sim \text{Po}(4.5)$.

The test value is $x = 2$, so, at the 5% level, reject H_0 if $\text{P}(X \leqslant 2) < 0.05$.

Since λ is small, the method is similar to the small sample binomial test.

$$\text{P}(X \leqslant 2) = \text{e}^{-4.5} + 4.5\text{e}^{-4.5} + \frac{4.5^2}{2!}\text{e}^{-4.5} = \text{e}^{-4.5}\left(1 + 4.5 + \frac{4.5^2}{2!}\right) = 0.173\ldots$$

Since $\text{P}(X \leqslant 2) > 0.05$, the test value does not lie in the critical region and H_0 is not rejected. The office manager's claim that the mean is 4.5 is upheld.

Progress check

1 $X \sim \text{N}(\mu, 4)$. The mean of a random sample of 10 items from X is 20.9.
Test, at the 5% level, the hypotheses H_0: $\mu = 20$, H_1: $\mu > 20$.

2 The heights of a particular plant have mean μ and variance σ^2 (both unknown). The heights H, in centimetres, of a random sample of 100 plants are summarised as follows: $\Sigma h = 2941$, $\Sigma h^2 = 88\,502$.

(a) Find unbiased estimates for μ and σ.

(b) Does the sample data support the claim that the mean height is less than 30 cm? Perform a hypothesis test at the 10% level and state your hypotheses.

3 A random observation, x is taken from $X \sim \text{B}(n, p)$.
Test, at the 5% level, the null hypothesis that $p = 0.4$, against the alternative hypothesis that $p < 0.4$.

Hint:
In (a) use small sample binomial test.
In (b) use large sample binomial test.

(a) if $n = 10$ and $x = 1$,

(b) if $n = 100$ and $x = 32$.

4 A single observation is taken from a Poisson distribution with mean λ and used to test the null hypothesis $\lambda = 7$ against the alternative hypothesis $\lambda \neq 7$.
What is the conclusion if the hypothesis test is carried out at the 10% level and the observation is

Cumulative Poisson tables are needed for (b).

(a) 3

(b) 13?

1 $z = 1.423$, do not reject H_0.

2 (a) 29.41, 4.503 (b) H_0: $\mu = 30$, H_1: $\mu < 30$; $z = -1.310$, lies in critical region so reject H_0 and uphold claim.

3 (a) $\text{P}(X \leqslant 1) = 0.0464 < 0.05$, reject H_0, conclude $p < 0.4$.

(b) $z = -1.531$, $z > -1.645$, do not reject H_0, conclude that p could be 0.4.

4 (a) $\text{P}(X \leqslant 3) > 0.05$, H_0 is not rejected. (b) $\text{P}(X \geqslant 13) < 0.05$, H_0 is rejected.

2.5 Hypothesis tests 2

After studying this section you should be able to:

LEARNING SUMMARY

- *use the t-distribution to obtain confidence intervals for the mean of a normal population with unknown variance, based on a small sample*
- *perform t-tests for the mean and difference between means of normal populations with unknown variance, using small samples*
- *perform tests for the product moment correlation coefficient and Spearman's rank correlation coefficient*

The t-distribution

AQA S4
OCR S3
NICCEA S3

If you take a small sample from a normal population with unknown mean μ and unknown variance, σ^2, the t-distribution is needed to obtain confidence intervals for μ and perform significance tests.

For large values of ν, the t-distribution approaches a normal distribution.

The **t-distribution** has a similar shape bell shape to the normal distribution.

It has one parameter, ν, known as the number of **degrees of freedom**.

For a particular value of ν, the appropriate t-distribution is denoted by $t(\nu)$. Tables give values of t for which $P(T < t) = p$, for various values of p, usually $p = 0.75$, 0.90, 0.95, 0.975, 0.99.

Confidence intervals 2

AQA S4
OCR S3
NICCEA S3

For small samples, size n, taken from $X \sim N(\mu, \sigma^2)$, with σ^2 unknown:

σ^2 is estimated using $\hat{\sigma}^2$ (see p. 56).

The critical t-value used in the confidence interval is found from t-distribution tables, with $\nu = n - 1$.

The 95% confidence limits for μ are $\bar{x} \pm t\dfrac{\hat{\sigma}}{\sqrt{n}}$,

where $(-t, t)$ encloses the central 95% of the $t(n-1)$ distribution.

The 95% confidence interval for μ is $\left(\bar{x} - t\dfrac{\hat{\sigma}}{\sqrt{n}},\ \bar{x} + t\dfrac{\hat{\sigma}}{\sqrt{n}}\right)$.

Example

The population must follow a normal distribution.

A random sample of 8 observations from $X \sim N(\mu, \sigma^2)$ gave $\Sigma x = 36.4$, $\Sigma x^2 = 188.08$

(a) Find an unbiased estimate of σ.
(b) Find a 95% confidence interval for μ.

(a) $$\hat{\sigma}^2 = \frac{1}{n-1}\left(\Sigma x^2 - \frac{(\Sigma x)^2}{n}\right)$$

$$= \frac{1}{7}\left(188.08 - \frac{(36.4)^2}{8}\right) = 3.208\ \ldots$$

so $\hat{\sigma} = \sqrt{3.208\ \ldots} = 1.791\ \ldots$

You need to find t such that $P(-t < T < t) = 0.95$, i.e. $P(T < t) = 0.975$.

(b) Since $\nu = 7$, use the $t(7)$ distribution.
Since the confidence interval is symmetric (two-tailed), find $\nu = 7$, $p = 0.975$ (2.5% in the upper tail). This gives critical value 2.365.

The 95% confidence limits for μ are $\bar{x} \pm t\dfrac{\hat{\sigma}}{\sqrt{n}} = 4.55 \pm 2.365\left(\dfrac{1.791\ \ldots}{\sqrt{8}}\right)$

$= 4.55 \pm 1.497\ \ldots$

The 95% confidence interval $= (4.55 - 1.497\ \ldots,\ 4.55 + 1.497\ \ldots)$
$= (3.05, 6.05)$ (2 d.p.)

t-tests

AQA	S4
OCR	S3
NICCEA	S3

t-test for the mean

When testing the mean from a normal population with unknown variance, when the sample size is small, the test statistic T follows a $t(n-1)$ distribution.

The distribution of X must be normal.

$\hat{\sigma}$ is the unbiased estimate of σ.

KEY POINT

$$\text{Test statistic } T = \frac{\bar{X}-\mu}{\hat{\sigma}/\sqrt{n}} \sim t(n-1)$$

Example

Using the data given in the preceding example:
test at the 5% level the claim that μ is greater than 4.

The null hypothesis is $\mu = 4$, even though you are asked to test the claim that $\mu > 4$.

H_0: $\mu = 4$, H_1: $\mu > 4$.

You want t such that $P(T < t) = 0.95$.

Using t-tables for $t(7)$,
the critical value for $\nu = 7$, $p = 0.95$ (one-tailed test, 5% in the upper tail) is 1.895 so reject H_0 if $t > 1.895$.

$$\bar{x} = \frac{\sum x}{n} = \frac{36.4}{8} = 4.55,$$

$$\Rightarrow t = \frac{\bar{x}-\mu}{\hat{\sigma}/\sqrt{n}} = \frac{4.55-4}{1.791/\sqrt{8}} = 0.868\ \ldots$$

Since $t < 1.895$, the test value does not lie in the critical region. There is no evidence to support the claim that μ is greater than 4.

t-test for the difference between means

If σ^2 is known, then the z-test can be used (see p. 63).

To test the difference between means of two independent normal populations with a common population variance σ^2 (unknown) using two small unmatched samples:

The number of degrees of freedom is $(n_1-1)+(n_2-1) = n_1+n_2-2$.

Check the version used in your examination formulae booklet.

KEY POINT

$$\text{Test statistic } T = \frac{\bar{X}_1-\bar{X}_2-(\mu_1-\mu_2)}{\hat{\sigma}\sqrt{\frac{1}{n_1}+\frac{1}{n_2}}} \sim t(n_1+n_2-2)$$

where $\hat{\sigma}^2 = \dfrac{n_1 s_1^2 + n_2 s_2^2}{n_1+n_2-2}$ — s_1^2 and s_2^2 are the sample variances

or $\hat{\sigma}^2 = \dfrac{(n_1-1)\hat{\sigma}_1^2 + (n_2-1)\hat{\sigma}_2^2}{n_1+n_2-2}$ — $\hat{\sigma}_1^2$ and $\hat{\sigma}_2^2$ are the unbiased estimates of σ_1^2 and σ_2^2

Test for correlation coefficient

EDEXCEL S3
NICCEA S2

Key points from AS

- **Correlation and regression** *Revise AS page 104*

The procedure is the same for both the product-moment and Spearman's rank test.

In the AS course you may have calculated the **product-moment correlation coefficient** and **Spearman's rank correlation coefficient.**

Both coefficients take values between –1 and 1. A hypothesis test enables you to test the strength of the correlation. For example, is a coefficient of 0.6 high enough to say that that there is positive correlation?

The following notation is often used.
Product-moment coefficient: population value ρ, sample value r
Spearman's rank coefficient: population value ρ_s, sample value r_s

Stages in the hypothesis test for correlation coefficient

Make the **null hypothesis, H_0** that the correlation coefficient is zero.

The test is one-tailed or two-tailed, depending on the **alternative hypothesis H_1**.
H_0: $\rho = 0$, H_1: $\rho > 0$ (one-tailed, upper tail test for positive correlation).
H_0: $\rho = 0$, H_1: $\rho < 0$ (one-tailed, lower tail test for negative correlation).
H_0: $\rho = 0$, H_1: $\rho \neq 0$ (two-tailed test for correlation, positive or negative).

Decide on the **significance level**, such as 10%, 5% or 1%. This defines the **critical values** for the **rejection region.**

The critical values can be obtained from tables, for **example**, at the 5% level:

For a one-tailed test, find the value in the column headed 0.05 (5% in the tail). This is the critical value for an upper tail test. Take the negative of it for a lower tail test.

For a two-tailed test, look for the column headed 0.025 (2.5% in each tail). Take that value for the upper critical value and the negative of it for the lower critical value.

There are two critical values in the two-tailed test, one in the upper tail and one in the lower tail.

Calculate the sample correlation coefficient and compare it with the critical value.
If sample value > critical value, reject H_0 (in the upper tail).
If sample value < critical value, reject H_0 (in the lower tail).

Make your decision and relate it to the situation being investigated.

Example

The 10 students in a mathematics class obtained these results in their mock and actual examinations.
Find the product moment correlation coefficient and test, at the 1% level, whether there is positive correlation between the results in the two examinations.

Candidate	A	B	C	D	E	F	G	H	I	J
Mock result	45	62	85	59	27	73	74	51	72	71
Actual result	54	67	72	73	31	79	63	61	80	74

Take care: If you use a calculator and obtain the wrong answer, you will get no marks! If you use the formula, you may get some method marks.
Practise using the format in your examination booklet.

From the calculator in LR mode: $r = 0.8572$.

H_0: $\rho = 0$, H_1: $\rho > 0$, where the population correlation coefficient is ρ.

From tables, at 1% level, the critical value is 0.6851, so H_0 is rejected if $r > 0.6851$.

Since $r = 0.8572 > 0.6851$, H_0 is rejected. There is evidence, at the 1% level, of positive correlation between the results in the two examinations.

Progress check

1 Using the data for the examination results in the above example:

(a) Calculate Spearman's coefficient of rank correlation.

(b) Test, at the 1% level, whether there is evidence of agreement between the rankings of the sets of marks of the candidates.

2 It is known that the times, in seconds, taken to perform a particular task are normally distributed with mean μ and variance σ^2.
The times, in seconds, taken by 10 volunteers are as follows:
32.9, 27.6, 33.0, 35.7, 39.2, 26.8, 34.3, 31.1, 38.9, 28.7

(a) Find an unbiased estimate of σ.

(b) Find a 99% confidence interval for μ.

(c) Find the width of the 95% confidence interval.

(d) Test at the 5% level the null hypothesis that $\mu = 35.5$, against the alternative hypothesis that $\mu < 35.5$.

3 $X \sim N(\mu_1, \sigma^2)$, $Y \sim N(\mu_2, \sigma^2)$ and X and Y are independent.
A sample of 10 observations from X has mean 15 and standard deviation 3.
A sample of 8 observations from Y has mean 20 and standard deviation 4.

(a) Obtain an unbiased estimate for σ.

(b) Perform a test, at the 5% level, to test whether the difference between the means is significant.

1 (a) 0.6121 (b) No evidence of agreement between the rankings.
2 (a) 4.37 (b) (28.3, 37.3) (c) 6.25
(d) $t = -1.941$, reject H_0 since $t < -1.833$; mean time is less than 35.5 seconds.
3 (a) 3.69
(b) $t = -2.856$, reject H_0 since $t < -2.12$, the difference is significant

2.6 χ^2 tests

After studying this section you should be able to:

- perform χ^2 goodness-of-fit tests
- perform χ^2 tests for association (independence) using contingency tables

LEARNING SUMMARY

χ^2 goodness-of-fit tests

EDEXCEL S3
OCR S3
NICCEA S3

A χ^2 goodness-of-fit test is used to test how well a statistical distribution models a given set of data.

Stages of a χ^2 goodness-of-fit test:

- State the null hypothesis, H_0 (that the data follow a particular distribution) and the alternative hypothesis, H_1 (that they do not).
- Calculate the frequencies that you would expect if the data do follow that distribution (these are the expected frequencies E).
  The expected frequencies must be greater than 5, so combine any adjacent classes if necessary. Also combine corresponding observed frequencies.
- Denoting the observed or actual frequencies by O, calculate the value of the test statistic χ^2 where $\chi^2 = \sum \frac{(O-E)^2}{E}$.
- Work out the number of degrees of freedom, ν. This is the number of independent variables used in calculating χ^2 (see list below).
- Decide on the significance level of the test and look up the critical value in χ^2 tables.
- Make your conclusion:
  If $\chi^2 >$ critical value, reject H_0 and conclude that the data cannot be modelled by the distribution specified in the null hypothesis.
  If $\chi^2 <$ critical value, do not reject H_0 and conclude that the distribution specified in H_0 provides a good model for the data.

Sometimes the notation f_e is used for the expected frequency and f_o for the observed frequency. Then $\chi^2 = \sum \frac{(f_o - f_e)^2}{f_e}$

Degrees of freedom, ν

The number of classes is denoted by n

ν = number of classes – number of restrictions

For a uniform distribution or one in a given ratio, $\nu = n - 1$

For a binomial distribution:
If p is known, $\nu = n - 1$
If p is unknown and estimated from observed data, $\nu = n - 2$

To estimate p, use $np = \bar{x}$

For a Poisson distribution:
If λ is known, $\nu = n - 1$
If λ is estimated from observed data, $\nu = n - 2$

To estimate λ, use $\lambda = \bar{x}$

For a normal distribution:
If μ and σ^2 are known, $\nu = n - 1$
If μ and σ^2 are estimated from the observed data, $\nu = n - 3$

Use $\hat{\mu}$ and $\hat{\sigma}^2$

Example

Perform a χ^2 goodness-of-fit test, at the 5% significance level, to test whether the following data could have come form a discrete uniform distribution.

x	0	1	2	3	4	
f	10	17	11	6	16	$\Sigma f = 60$

The expected frequencies have to be calculated before the rejection rule is set, because if $E < 5$, categories have to be combined and this will affect the value of ν.

H_0: The distribution is uniform
H_1: The distribution is not uniform

If the distribution is uniform, $E = \frac{60}{5} = 12$ for each category

There are 5 categories so $\nu = 5 - 1 = 4$ and the $\chi^2(4)$ distribution is considered. At the 5% level, the critical value (from tables) is 9.488, so H_0 is rejected if $\chi^2 > 9.488$.

It is useful to write the values in a table for clarity.

x	O	E	$\frac{(O-E)^2}{E}$
0	10	12	0.333 ...
1	17	12	2.0833 ...
2	11	12	0.0833 ...
3	6	12	3 ...
4	16	12	1.333 ...
	$\Sigma O = 60$	$\Sigma E = 60$	$\chi^2 = 6.833$ ...

$$\chi^2 = \sum \frac{(O-E)^2}{E}$$
$$= 6.833 \text{ (3 d.p.)}$$

Do a running total on your calculator.

Since $\chi^2 < 9.488$, the test value does not lie in the critical region and H_0 is not rejected. The data could have come from a uniform distribution.

χ^2 tests for association using contingency tables

AQA	S4
EDEXCEL	S3
OCR	S3
NICCEA	S3

This χ^2 test is used to test whether two factors are independent or whether there is an association between them. The procedure is similar to that for the χ^2 goodness-of-fit test.

The null hypothesis, H_0, is that the two factors are independent.
The alternative hypothesis, H_1, is that there is an association between them.

	B			
	x	x	x	x
A	x	x	x	x
	x	x	x	x

This is a 3 by 4 table

Data are in the form of a **contingency table**, which is an array displaying data relating to the two factors.

The size of the table is described by the number of rows and the number of columns. An h by k table has h rows and k columns.

The expected frequency for a particular cell is calculated as follows:

Expected frequencies should be greater than 5, so it may be necessary to combine adjacent rows or columns. This must be done sensibly, with regard to the categories.

$$E = \frac{(\text{row total}) \times (\text{column total})}{\text{grand total}}$$

The test statistic is $\chi^2 = \sum \frac{(O-E)^2}{E}$.

This is the same formula as for a goodness-of-fit test.

The number of degrees of freedom, ν, depends on the size of the contingency table. For an h by k table, $\nu = (h-1)(k-1)$.

For a 2 by 2 contingency table, $\nu = 1$ and it is advisable to use Yates' continuity correction. In this case $\chi^2 = \sum \frac{(|O-E|-0.5)^2}{E}$.

Example

Test, at the 10% level, the null hypothesis that size and colour are independent:

This is a 3 by 2 contingency table

		Colour		
		Pink	**White**	**Totals**
Size	**Small**	12	23	35
	Medium	16	20	36
	Large	22	17	39
	Totals	50	60	110

H_0: Size and colour are independent;
H_1: Size and colour are not independent

Expected frequencies (to 1 decimal place).

Small pink: $E = \dfrac{(\text{row total}) \times (\text{column total})}{\text{grand total}} = \dfrac{35 \times 50}{110} = 15.9$

Only two expected frequencies need to be calculated using the formula. All the rest can then be found by ensuring that row or column totals agree. This confirms that $\nu = 2$.

Medium pink: $E = \dfrac{36 \times 50}{110} = 16.4$

Large pink: $E = 50 - (15.9 + 16.4) = 17.7$ (column totals argree).
Small white: $E = 35 - 15.9 = 19.1$ (row totals agree).
Medium white: $E = 36 - 16.4 = 19.6$ (row totals agree).
Large white: $E = 39 - 17.7 = 21.3$ (row totals agree).

$\nu = (3-1)(2-1) = 2 \times 1 = 2$. Consider the $\chi^2(2)$ distribution. At the 10% level, the critical value (from tables) is 4.605, so H_0 is rejected if $\chi^2 > 4.605$

O	E	$\dfrac{(O-E)^2}{E}$
12	15.9	0.956 ...
16	16.4	0.0097 ...
22	17.7	1.044 ...
23	19.1	0.796 ...
20	19.6	0.00816 ...
17	21.3	0.868 ...
$\Sigma O = 160$	$\Sigma E = 160$	$\chi^2 = 3.6835$ ...

$\chi^2 = \sum \dfrac{(O-E)^2}{E} = 3.68$

Since $\chi^2 < 4.605$, H_0 is not rejected.
Size and colour are independent.

Progress check

1 Carry out a goodness-of-fit test at the 5% level to test whether the following data could have come from a binomial distribution with $p = 0.25$.

x	0	1	2	3
f	199	220	65	14

2 Test, at the 5% level, whether there is an association between the two characteristics in this contingency table.

	Characteristic 2			Totals
Characteristic 1	90	35	85	210
	30	30	60	120
	30	15	25	70
Totals	150	80	170	400

1 $\nu = 3$, $E = 210.09, 210.09, 70.03, 7.78$, $X^2 = 6.39 < 7.815$, yes it could.
2 $\nu = 4$, $E = 78.75, 42, 89.25, 45, 24, 51, 26.25, 14, 29.75$, $X^2 = 12.43 > 9.488$, there is an association.

Sample questions and model answers

1

The continuous random variable X has probability density function $f(x)$ given by

$$f(x) = \begin{cases} kx(2-x) & 0 \leqslant x \leqslant 2 \\ 0 & \text{otherwise} \end{cases}$$

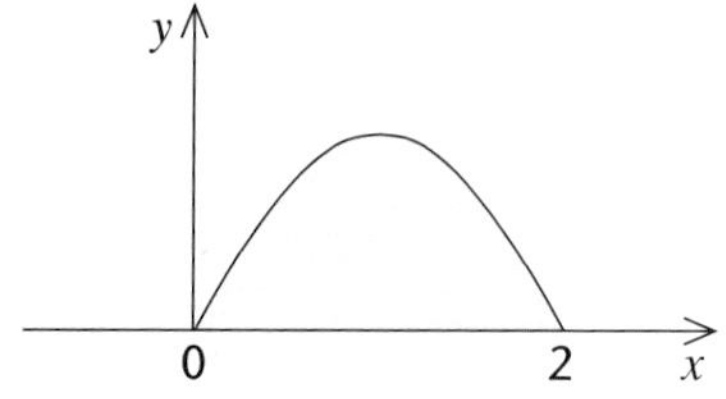

where k is a constant.

(a) Show that $k = \frac{3}{4}$ and find the exact value of $P(X > 1\frac{1}{2})$.

(b) Find $E(X)$ and $E(X^2)$.

(c) Show that the standard deviation, σ, is 0.447 correct to three decimal places.

The total area under the curve is 1.

(a) $$\int_0^2 f(x)dx = 1$$

As k is a constant, it can be taken outside the integration.

$$\Rightarrow \quad k\int_0^2 (2x - x^2)dx = 1$$

$$\Rightarrow \quad k\left[x^2 - \frac{x^3}{3}\right]_0^2 = 1$$

$$\Rightarrow \quad k(4 - \tfrac{8}{3} - 0) = 1$$

$$\Rightarrow \quad k = \tfrac{3}{4}$$

$$f(x) = \tfrac{3}{4}(2x - x^2)$$

$$P(X > 1\tfrac{1}{2}) = \int_{1\frac{1}{2}}^2 f(x)dx$$

$$= \tfrac{3}{4}\int_{1\frac{1}{2}}^2 (2x - x^2)dx$$

$$= \tfrac{3}{4}\left[x^2 - \frac{x^3}{3}\right]_{1\frac{1}{2}}^2$$

Substitute the limits carefully, working in fractions to obtain the exact answer.

$$= \tfrac{3}{4}(4 - \tfrac{8}{3} - (\tfrac{9}{4} - \tfrac{9}{8}))$$

$$= \tfrac{5}{32}$$

If you do not spot that $f(x)$ is symmetrical about $x = 1$, find $E(X)$ by integration.

(b) By symmetry, $E(X) = 1$

$$E(X^2) = \int_0^2 x^2 f(x)dx$$

$$= \tfrac{3}{4}\int_0^2 (2x^3 - x^4)dx$$

$$= \tfrac{3}{4}\left[\frac{2x^4}{4} - \frac{x^5}{5}\right]_0^2$$

$$= \tfrac{3}{4}(8 - \tfrac{32}{5})$$

$$= 1.2$$

(c) $\text{Var}(X) = E(X^2) - [E(X)]^2$

$[E(X)]^2$ is often written $E^2(X)$.

$$= 1.2 - 1$$

$$= 0.2$$

$$\sigma = \sqrt{0.2} = 0.447 \text{ (3 d.p.)}$$

Sample questions and model answers (continued)

2

In an advanced level examination taken by a large number of candidates, the marks were distributed normally with mean mark 68.7 and standard deviation 5.4.

A random sample of 100 scripts is taken and their mean mark denoted by $\overline{X}$.

(a) State the distribution of $\overline{X}$.

(b) Find the probability that the mean mark of the 100 scripts is between 68 and 70, giving your answer correct to 2 significant figures.

Let X be the examination mark, so $X \sim N(68.7, 5.4^2)$

(a) For samples of size 100,

$$\overline{X} \sim N\left(68.7, \frac{5.4^2}{100}\right), \text{ i.e. } \overline{X} \sim N(68.7, 0.2916)$$

(b) $P(68 < \overline{X} < 70) = P\left(\frac{68 - 68.7}{\sqrt{0.2916}} < Z < \frac{70 - 68.7}{\sqrt{0.2916}}\right)$

$= P(-1.296 < Z < 2.407)$

$= \Phi(1.296) + \Phi(2.407) - 1$

$= 0.9026 + 0.9919 - 1 = 0.89$ (2 s.f.)

If the degree of approximation is not specified in the question or your examination rubric, then approximate sensibly.

3

The masses of a particular brand of buns are normally distributed with mean 50 g and standard deviation 3.4 g. They are sold in small packs of 6 or large packs of 12.

(a) Find the probability that the mass of the buns in a small pack is less than 285 g.

(b) Find the probability that the buns in two small packs weigh at least 10 g more than the buns in a large pack.

Always define the random variables carefully as this will help you to keep track of the distribution being used.

Let X be the mass of a bun. Then $X \sim N(50, 3.4^2)$.

Note that $S = X_1 + X_2 + \ldots + X_6$ (sum) $S \neq 6X$ (multiple)

(a) Let S be the mass of a small pack, where $S = X_1 + X_2 + \ldots + X_6$

$E(S) = 6E(X) = 300$ and $\text{Var}(S) = 6\,\text{Var}(X) = 6 \times 3.4^2 = 69.36$

$\therefore\ S \sim N(300, 69.36)$

$$P(S < 285) = P\left(Z < \frac{285 - 300}{\sqrt{69.36}}\right) = P(Z < -1.801) = 0.036 \text{ (2 s.f.)}$$

(b) Let L be the mass of a large pack, where $L = X_1 + X_2 + \ldots + X_{12}$

$E(L) = 12E(X) = 600$ and $\text{Var}(L) = 12\,\text{Var}(X) = 12 \times 3.4^2 = 138.72$

$\therefore\ L \sim N(600, 138.72)$.

$P(S_1 + S_2 \geqslant L + 10) = P(S_1 + S_2 - L \geqslant 10)$

Let $D = S_1 + S_2 - L$

Remember the + sign in the variance.

$E(D) = 300 + 300 - 600 = 0$ and $\text{Var}(D) = 69.36 + 69.36 + 138.72 = 277.44$

$\therefore\ D \sim N(0, 277.44)$

$$P(D \geqslant 10) = P\left(Z \geqslant \frac{10 - 0}{\sqrt{277.44}}\right) = P(Z \geqslant 0.6003\ldots) = 0.27 \text{ (2 s.f.)}$$

Sample questions and model answers (continued)

4

The lengths of a population of snakes are normally distributed with standard deviation 8 cm and unknown mean μ cm.

A random sample of 10 snakes is taken and their mean length calculated.

When a significance test, at the 10% level, is performed, the hypothesis that μ is 37.5 is rejected in favour of the hypothesis that it is greater than 37.5 cm.

(a) What can be said about the value of the sample mean?

(b) Explain briefly what is meant, in the context of this question, by a Type I error, and state the probability of making a Type I error.

(c) Find the probability of making a Type II error when $\mu = 41.1$.

(a) If X is the length, in centimetres, then $X \sim N(\mu, 8^2)$.

H_0: $\mu = 37.5$

H_1: $\mu > 37.5$

For the significance test, the sampling distribution of means is considered.

If $\mu = 37.5$, then $\bar{X} \sim N\left(37.5, \frac{8^2}{10}\right)$.

From standard normal tables, $\Phi(z) = 0.9 \Rightarrow z = 1.282$

The critical value for a one-tailed (upper tail) at the 10% level is 1.282.

H_0 is rejected if z is in the upper tail 10% of the distribution.

Since H_0 is rejected, $z > 1.282$,

Test statistic is $Z = \frac{\bar{X} - \mu}{\sigma/\sqrt{n}}$

i.e. $$\frac{\bar{x} - 37.5}{8/\sqrt{10}} > 1.282$$

$$\bar{x} > 40.74$$

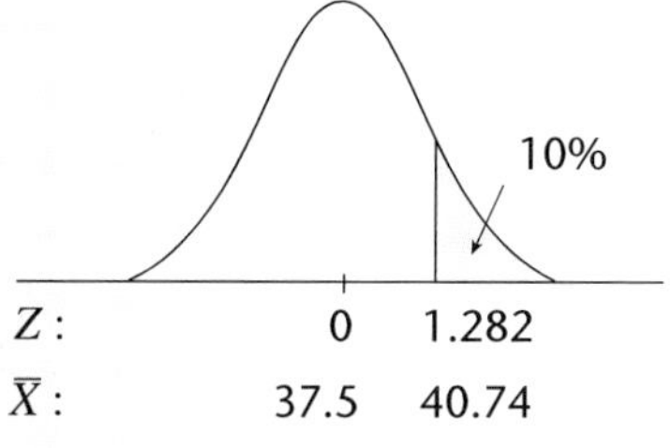

(b) A Type I error is made when the null hypothesis, $\mu = 37.5$, is rejected, when the mean population length is in fact 37.5.

P(Type I error) = 10%

The probability of making a Type I error is the same as the level of significance of the test.

(c) H_0: $\mu = 37.5$

H_1: $\mu = 41.1$

P(Type II error) = P(H_0 is accepted | H_1 is true)

From (a), H_0 is accepted if $\bar{x} < 40.74$.

If H_1 is true, $\bar{X} \sim N\left(41.1, \frac{8^2}{10}\right)$

$$\text{P(Type II error)} = P\left(\bar{X} < 40.74 \middle| \bar{X} \sim N\left(41.1, \frac{8^2}{10}\right)\right)$$

$$= P\left(Z < \frac{40.74 - 41.1}{8/\sqrt{10}}\right)$$

$$= P(Z < -0.142)$$

$$= 1 - 0.5565$$

$$= 0.44 \text{ (2 s.f.)}$$

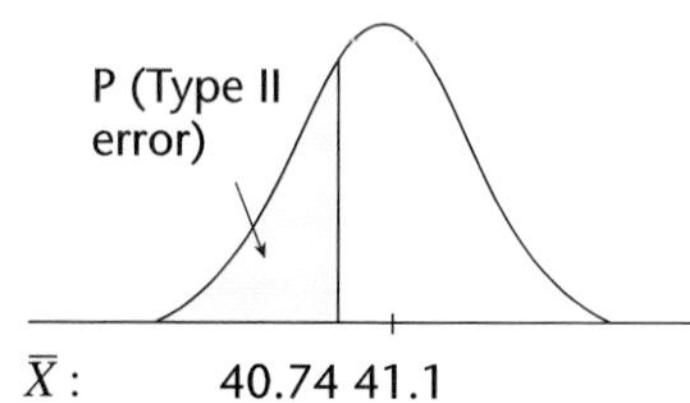

Practice examination questions

1 It is given that $X \sim \mathrm{N}(\mu, 16)$.
The null hypothesis $\mu = 20$ is to be tested against the alternative hypothesis $\mu \neq 20$.

The mean of a random sample of 10 observations from X is 17.2.

Test at the 5% level whether this provides evidence that μ is not 20.

2 Two hundred randomly chosen sixth form students were asked whether they had a Saturday job and 132 replied yes.

(a) Calculate an approximate 99% confidence interval for p, the population proportion of sixth form students who have a Saturday job. Give the end-points correct to 2 decimal places.

(b) Give a reason why p may not be contained in this interval.

(c) State the effect on the width of the confidence interval if the confidence level is decreased.

3 The continuous random variable X is distributed uniformly in the interval $2 \leqslant x \leqslant 12$.

(a) Find the probability density function, $f(x)$.

(b) Calculate the probability that X lies within one standard deviation of the mean, giving your answer correct to 2 decimal places.

4 The random variable X has a binomial distribution with $n = 10$ and $p = 0.4$.
The mean of 60 random observations of X is $\overline{X}$.

Find $\mathrm{P}(\overline{X} < 3.5)$, explaining what part the central limit theorem has played in your answer.

5 A beetle infestation is discovered in the trees in a small copse. If more than 35% of the trees are infected it is impractical to treat the trees with chemicals and they will have to be felled.

(a) The representative from the ministry checked a random sample of 10 trees and found that 6 were infected. Stating any assumptions, use a hypothesis test, at the 10% significance level, to decide whether the infestation should be treated with chemicals or the trees in the copse felled.

(b) The estate manager checked a random sample of 30 trees and found that 14 were infected. Using a suitable approximation, carry out a significance test, again at the 10% level. On the basis of this sample, how should the infestation be dealt with?

Practice examination questions (continued)

6 The random variable X has a normal distribution with mean μ and variance σ^2. In order to set up a confidence interval for μ, a random sample of 9 observations of X is taken.

The results are summarised as follows:

$$\Sigma x = 255.6 \qquad \Sigma x^2 = 7372.26$$

(a) If $\sigma = 4$, calculate a 90% confidence interval for μ.

(b) If σ is unknown,

(i) calculate an unbiased estimate for σ,

(ii) calculate a 95% confidence interval for μ.

7 The continuous random variable X has cumulative distribution function given by

$$F(x) = \begin{cases} 0 & x \leqslant 0 \\ 2cx & 0 \leqslant x \leqslant 1 \\ c(4x - x^2 - 1) & 1 \leqslant x \leqslant 2 \\ 1 & x \geqslant 2 \end{cases}$$

(a) Show that $c = \frac{1}{3}$.

(b) Show that $P(X > 1\frac{1}{2}) = \frac{1}{12}$.

(c) Find the probability function $f(x)$.

(d) Calculate $E(X)$.

8 In a cross-pollination experiment, 160 seeds germinated with the following results:

Category	*Characteristics*	*Number*
A	Tall with red flowers	107
B	Dwarf with red flowers	24
C	Tall with white flowers	23
D	Dwarf with white flowers	6

Perform a χ^2 goodness-of-fit test, at the 5% significance level, to find out whether the results support the theory that the categories A : B : C : D occur in the ratio 9 : 3 : 3 : 1.

9 The management of a supermarket is investigating the average amounts spent per visit of customers to its stores. During a particular week, a random sample of 60 customers in its inner-city store and 75 customers in its out-of-town store were surveyed.

The mean amount spent in the inner-city store was £19.27 with standard deviation £4.27 and the mean amount in the out-of-town store was £21.04 with standard deviation £6.89.

(a) The amounts spent in each store have a common population standard deviation σ. Calculate the unbiased estimate of σ.

(b) Test whether the mean amounts differ in the two stores. Use a 10% level of significance and state your hypotheses clearly.

Practice examination questions (continued)

10 The marks awarded by Judge 1 and Judge 2 to the finalists in the 'Best of Breed' category at the annual Dog Show were as follows:

Dog	A	B	C	D	E	F	G	H
Judge 1	25	35	32	24	20	45	55	56
Judge 2	29	27	46	42	50	30	34	35

A mathematical dog owner was unhappy with the marking and carried out a significance test at the 5% level on Spearman's coefficient of rank correlation.

Did this provide evidence that there was strong disagreement between the rankings of the two judges?

11 An opinion poll is conducted two weeks before an election for a union leader. In a randomly selected sample of 500 members of the union in London, 236 said that they intended to vote for candidate A. The corresponding number in Manchester was 156 out of 300.

(a) From the sample data, calculate an unbiased estimate of the proportion of the voting population who intend to vote for Candidate A.

(b) Test, at the 10% level of significance, whether there is a significant difference between the proportions in the two cities who intend to vote for Candidate A.

12 A psychologist is investigating whether or not there is any association between whether a person is left- or right-handed and the ability to complete a task involving manual dexterity in a given time. A sample of 200 people gave the following results:

	Completed task	*Did not complete task*
Right-handed	83	67
Left-handed	37	13

Stating your hypotheses clearly, test, at the 5% level, whether or not there is any evidence of an association between right- or left-handedness and the ability to complete the task in the given time.

13 The probability that a person suffers from a particular health condition is 0.01.

(a) Using a suitable approximation, find the probability that in a randomly chosen sample of 80 people, fewer than 3 suffer from the health condition.

(b) Find the minimum sample size in order that the probability of including at least one with the health condition is greater than 95%.

Chapter 3

Mechanics 2

The following topics are covered in this chapter:

- Projectiles
- Variable acceleration
- Equilibrium of a rigid body
- Centre of mass
- Collisions and impulse
- Uniform circular motion
- Work, energy and power

3.1 Projectiles

LEARNING SUMMARY

After studying this section you should be able to:

- model the motion of a projectile moving under constant acceleration
- understand the limitations of the model
- solve problems by considering, separately, the vertical and horizontal components of velocity
- derive and apply the trajectory formula

The model

AQA	M2
EDEXCEL	M2
OCR	M2
WJEC	M2, M5
NICCEA	M2

In the usual **particle** model for projectiles, air resistance is ignored and the only force taken to act on the particle is its weight which is constant. According to this model:

Check the value of g used on your exam paper. Make sure that you use the given value or you may lose marks.

KEY POINT

- The horizontal component of velocity remains constant throughout the motion.
- The vertical component of velocity is subject to a constant downward acceleration of g ms^{-2}.

The accuracy of the model, in predicting the motion of a projectile, depends on the extent to which the initial assumptions are satisfied.

For example, the model will provide accurate information about a projectile such as a stone or a dart over a short distance, but will give poor results for light objects that are affected more by air resistance.

Be careful about using formulae for time of flight, greatest height, range etc. without justification or, again, marks may be lost.

When using the model, a first step is usually to resolve the velocity of a projectile into horizontal and vertical components. The two parts are then treated separately.

A particle moving with speed v at angle θ to the horizontal has vertical component $v \sin \theta$ and horizontal component $v \cos \theta$.

The velocity is usually shown on a diagram as in Figure 1 but is used in equations as in Figure 2.

If you know the horizontal and vertical components of a velocity then you can combine them to find its magnitude (speed) and direction.

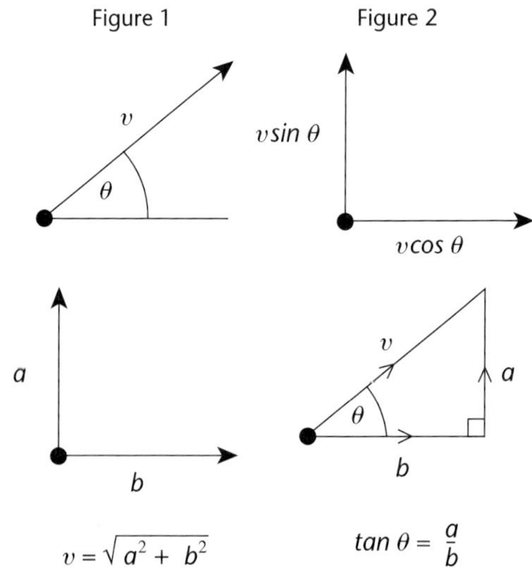

Upwards is taken to be positive so the acceleration is shown as negative.

Example

A stone is thrown with speed 15 ms^{-1} at an angle of 60° above the horizontal. Find its speed and direction after 1 second. Take $g = 9.8$ ms^{-2}.

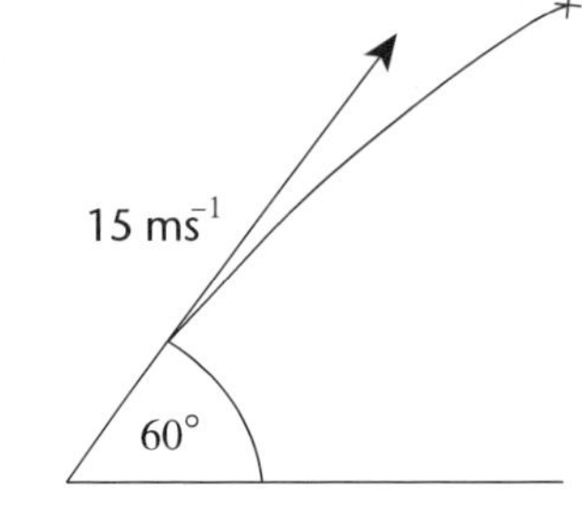

Vertically: using $v = u + at$

$$v_a = 15 \sin 60^0 - 9.8$$
$$= 3.190\ 38 \ldots$$

Horizontally: $v_b = 15 \cos 60^0$
$$= 7.5$$

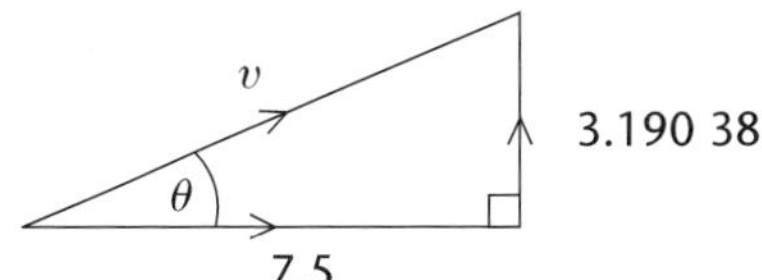

$v = \sqrt{7.5^2 + 3.190\ 38^2} = 8.15$ to 3 s.f.

$\tan\theta = \dfrac{3.19038}{7.5} \Rightarrow \theta = 23.0^0$ to 1 d.p.

After 1 second the stone has speed 8.15 ms^{-1} at 23^0 above the horizontal.

The trajectory formula

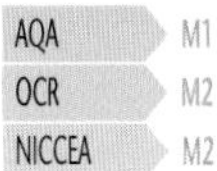

Taking the point of projection as the origin, the position of a projectile may be described in terms of (x, y) coordinates.

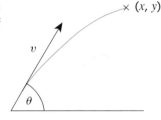

Horizontally: $x = v \cos\theta \times t \Rightarrow t = \dfrac{x}{v\cos\theta}$ (1)

Vertically: $y = v \sin\theta \times t - \frac{1}{2}gt^2$ (2)

Substituting for t in (2) gives: $y = v \sin\theta \times \dfrac{x}{v\cos\theta} - \dfrac{gx^2}{2v^2\cos^2\theta}$

This is the Cartesian equation of the path of the projectile. It is known as the trajectory formula.

so: $y = x\tan\theta - \dfrac{gx^2}{2v^2}(1 + \tan^2\theta)$

Note that: $\dfrac{1}{\cos^2\theta} \equiv \sec^2\theta \equiv (1 + \tan^2\theta)$

Example

A ball is kicked with speed 20 ms^{-1} at 30° above the horizontal towards a wall 5 m high. The wall is 15 m from the point where the ball is kicked. Take $g = 9.8$ ms^{-2}. Will the ball hit the wall or clear it? Consider the effect of any assumptions.

Using the formula: $y = x\tan\theta - \dfrac{gx^2}{2v^2}(1 + \tan^2\theta)$

$$y = 15\tan 30° - \frac{9.8 \times 15^2}{2 \times 20^2}(1 + \tan^2 30°)$$
$$= 4.99 \text{ m to 3 s.f.}$$

According to the model, the ball will hit the wall close to the top.
If the size of the ball is taken into account along with the effect of air resistance then it is clear that the ball cannot clear the wall.

Progress check

A small object is projected from ground level with speed 30 ms^{-1} at 40° above the horizontal. Take $g = 10$ ms^{-2}.

(a) Find the speed and direction of the object after 1.5 seconds.

(b) The object just clears a tree 25 m from the point of projection. How high is the tree?

(a) 23.4 ms^{-1}, 10.6° above horiz (b) 15.2 m

3.2 Variable acceleration

After studying this section you should be able to:

- *differentiate and integrate a vector with respect to time*
- *use differentiation and integration of vectors to solve problems involving variable acceleration*

LEARNING SUMMARY

Vector and scalar quantities

AQA M2
EDEXCEL M2
OCR M2
WJEC M2
NICCEA M2

Key points from AS

- **Scalars and vectors**
  Revise AS page 72

A **scalar** quantity has **magnitude** i.e. size, but not direction. Numbers are scalars and some other important examples are distance, speed, mass and time.

A **vector** quantity has both magnitude and direction. For example, distance in a specified direction is called displacement. Some other important examples are velocity, acceleration, force and momentum.

KEY POINT

The magnitude of a vector $\mathbf{r} = a\mathbf{i} + b\mathbf{j}$ is given by $r = \sqrt{a^2 + b^2}$

Differentiation of vectors

AQA M2
EDEXCEL M2
WJEC M2
NICCEA M2

A vector of the form $\mathbf{r} = f(t)\mathbf{i} + g(t)\mathbf{j}$ may be differentiated with respect to time to give $\dot{\mathbf{r}} = f'(t)\mathbf{i} + g'(t)\mathbf{j}$

For example, if $\mathbf{r} = 3t^2\mathbf{i} + \sqrt{t}\mathbf{j}$ then $\dot{\mathbf{r}} = 6t\mathbf{i} + \dfrac{1}{2\sqrt{t}}\mathbf{j}$.

Differentiate the coefficients of **i** and **j** with respect to t.

If r represents the position vector of some moving point at time t then $\dot{\mathbf{r}}$ represents the corresponding velocity at time t.

$\dot{\mathbf{r}}$ may be differentiated to give $\ddot{\mathbf{r}}$ which represents the acceleration of the moving point at time t.

Example

A particle P has position vector $\mathbf{r}$ metres relative to a fixed point O, at time t seconds, where $\mathbf{r} = (2t^2 + 5t)\mathbf{i} + t^3\mathbf{j}$ and $t \geqslant 0$.

Find:

(a) the velocity of P at time t seconds

(b) the speed of P when $t = 3$

(c) the time when P is moving parallel to the vector $\mathbf{i} + 3\mathbf{j}$.

(a) The velocity of P in metres per second at time t is given by $\dot{\mathbf{r}} = (4t + 5)\mathbf{i} + 3t^2\mathbf{j}$.

(b) When $t = 3$, $\dot{\mathbf{r}} = 17\mathbf{i} + 27\mathbf{j}$.

The speed of P is the magnitude of its velocity.

The speed of P when $t = 3$ is $\sqrt{17^2 + 27^2}$ ms^{-1} = 31.9 ms^{-1} to 3 s.f.

(c) When P is moving parallel to $\mathbf{i} + 3\mathbf{j}$ the component of velocity in the **j** direction must be 3 times the component in the **i** direction.

This gives:

$$3t^2 = 3(4t + 5)$$

so:

$$t^2 = 4t + 5$$

$$t^2 - 4t - 5 = 0$$

$$(t - 5)(t + 1) = 0$$

$$t = 5 \text{ or } t = -1$$

However, $t \geqslant 0$ so the only solution is $t = 5$.

Integration of vectors

AQA M2
EDEXCEL M2
WJEC M2
NICCEA M2

In general, $\int (f(t)\mathbf{i} + g(t)\mathbf{j})\mathrm{d}t = \left(\int f(t)\mathrm{d}t\right)\mathbf{i} + \left(\int g(t)\mathrm{d}t\right)\mathbf{j}$

Don't be put off by the complicated looking formula, just look at what it means.

So, to integrate a vector with respect to t, integrate the coefficients of $\mathbf{i}$ and $\mathbf{j}$ with respect to t. **Note that a constant of integration is needed and that this is itself a vector.**

Example

Find $\mathbf{v} = \int \mathbf{a}\,\mathrm{d}t$ given that $\mathbf{a} = 2t\mathbf{i} - 6t^2\mathbf{j}$ and that $\mathbf{v} = 3\mathbf{i} + \mathbf{j}$ when $t = 1$.

$$\mathbf{v} = \int (2t\mathbf{i} - 6t^2\mathbf{j})\mathrm{d}t = t^2\mathbf{i} - 2t^3\mathbf{j} + \mathbf{c}$$

When $t = 1$: $\mathbf{v} = \mathbf{i} - 2\mathbf{j} + \mathbf{c} = 3\mathbf{i} + \mathbf{j}$

so: $\mathbf{c} = 2\mathbf{i} + 3\mathbf{j}$

This gives: $\mathbf{v} = t^2\mathbf{i} - 2t^3\mathbf{j} + 2\mathbf{i} + 3\mathbf{j}$
or $\mathbf{v} = (t^2 + 2)\mathbf{i} + (3 - 2t^3)\mathbf{j}$

In the example, if $\mathbf{a}$ represents acceleration then $\mathbf{v}$ represents velocity.

KEY POINT

Vectors for position $\mathbf{r}$, velocity $\mathbf{v}$ and acceleration $\mathbf{a}$ are linked through differentiation and integration.

$\mathbf{v} = \dot{\mathbf{r}}$ $\qquad$ $\mathbf{v} = \int \mathbf{a}\,\mathrm{d}t$

$\mathbf{a} = \dot{\mathbf{v}} = \ddot{\mathbf{r}}$ $\qquad$ $\mathbf{r} = \int \mathbf{v}\,\mathrm{d}t$

differentiate →
$\mathbf{r}$ $\quad$ $\mathbf{v}$ $\quad$ $\mathbf{a}$
← integrate

Example

'Initially' means when $t = 0$.

A particle P, initially at rest, has acceleration $\mathbf{a} = 2t\mathbf{i} + 3\sqrt{t}\mathbf{j}$ ms^{-2} at time t seconds.

(a) Find the velocity of P at time t seconds.
(b) Show that P is moving parallel to $\mathbf{i} + \mathbf{j}$ when $t = 4$.

Note that the formulae for constant acceleration cannot be used here.

$3 \times t^{\frac{3}{2}} \times \frac{2}{3} = 2t^{\frac{3}{2}}$

(a) $\mathbf{v} = \int \mathbf{a}\,\mathrm{d}t = \int (2t\mathbf{i} + 3\sqrt{t}\mathbf{j})\mathrm{d}t$

$$= t^2\mathbf{i} + 2t^{\frac{3}{2}}\mathbf{j} + \mathbf{c}$$

When $t = 0$, $\mathbf{v} = \mathbf{0} \Rightarrow \mathbf{c} = \mathbf{0}$ so, $\mathbf{v} = t^2\mathbf{i} + 2t^{\frac{3}{2}}\mathbf{j}$

(b) When $t = 4$, $\mathbf{v} = 16\mathbf{i} + 16\mathbf{j}$
$= 16(\mathbf{i} + \mathbf{j})$ so P is moving parallel to $\mathbf{i} + \mathbf{j}$.

Progress check

1 An object has position vector $\mathbf{r} = (2t^3\mathbf{i} - 5t^2\mathbf{j})$ m at time t seconds. Find the acceleration of the object when $t = 3$.

2 A particle P, initially at the point with position vector $(-3\mathbf{i} + 5\mathbf{j})$ m, has velocity $\mathbf{v} = \sqrt{t}\mathbf{i} + 3\mathbf{j}$ at time t seconds. Find the position vector of P when $t = 4$.

1 $\ddot{\mathbf{r}} = 36\mathbf{i} - 10\mathbf{j}$ $\quad$ 2 $\frac{7}{3}\mathbf{i} + 17\mathbf{j}$

3.3 Equilibrium of a rigid body

After studying this section you should be able to:

- understand the conditions for equilibrium of a rigid body under the action of coplanar forces
- solve problems involving coplanar forces

LEARNING SUMMARY

Equilibrium

AQA	M2
EDEXCEL	M2
OCR	M2
WJEC	M3
NICCEA	M3

Forces that act in the same plane are coplanar.

You only need to establish this for *one* point and you can choose any point that is convenient.

You may need to show that a rigid body is in **equilibrium** under a system of **coplanar** forces *or* to *use* the fact that a rigid body is in equilibrium to find something about the forces acting on it.

To show that a rigid body is in equilibrium under the action of coplanar forces you need to establish that:

- the vector sum of the forces is zero
- the sum of the moments about some point is zero.

If a body is in equilibrium under some coplanar forces then you know that:

- the vector sum of the forces will be zero
- the sum of the moments about any point you choose will be zero.

> The conditions for equilibrium can be used to produce equations. The equations can then be solved to find the unknown values in a problem.
>
> KEY POINT

Example

A uniform plank AB of length 5 m and mass 20 kg rests horizontally on two supports, one at each end. A mass of 10 kg is positioned on the plank 1 m from B. Find the reaction at each of the supports. Take $g = 9.8$ ms^{-2}.

The first step is to draw a clearly labelled diagram.

The plank is uniform so its weight acts at its mid-point.

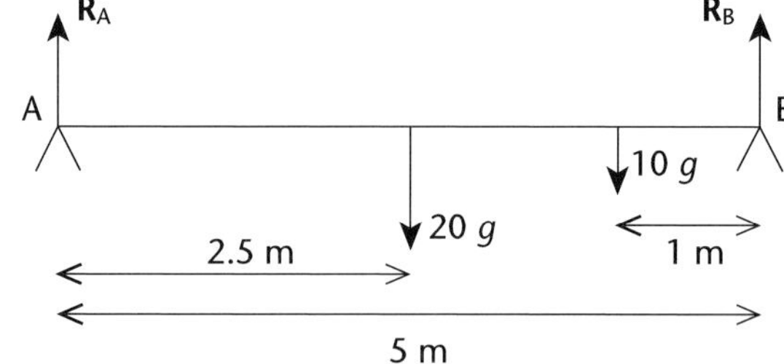

The vector sum of the forces is zero so the total upward force must equal the total downward force.

Vertically: $\mathbf{R}_A + \mathbf{R}_B = 20g + 10g$
$= 30g$

M(A) gives: $5\mathbf{R}_B = 20g \times 2.5 + 10g \times 4$
$= 90g$
$\mathbf{R}_B = 18g$
$= 176.4$ N (reaction at B)

Substitute for $\mathbf{R}_B$: $\mathbf{R}_A + 18g = 30g$
$\mathbf{R}_A = 12g$
$= 117.6$ N (reaction at A).

Leaning ladders

AQA	M2
EDEXCEL	M2
OCR	M2
WJEC	M3
NICCEA	M3

The situation where a ladder leans against a wall, possibly supporting a load at some point along its length, provides a rich source of questions on equilibrium. It requires forces and moments to be considered and includes the theory of friction.

Example

A uniform ladder of mass m kg rests against a smooth vertical wall with its lower end on rough horizontal ground. A man of mass $4m$ kg has climbed three-quarters of the way up the ladder but cannot go any higher or it will slip. The ladder makes an angle of 60° with the horizontal.
Find the coefficient of friction between the ladder and the ground.

Start by drawing a clearly labelled diagram.

The wall is smooth so there is no vertical force at the top of the ladder.

The length of the ladder has been written as $4l$ to simplify taking moments.

Friction acts at the bottom of the ladder to oppose the tendency for it to slip.

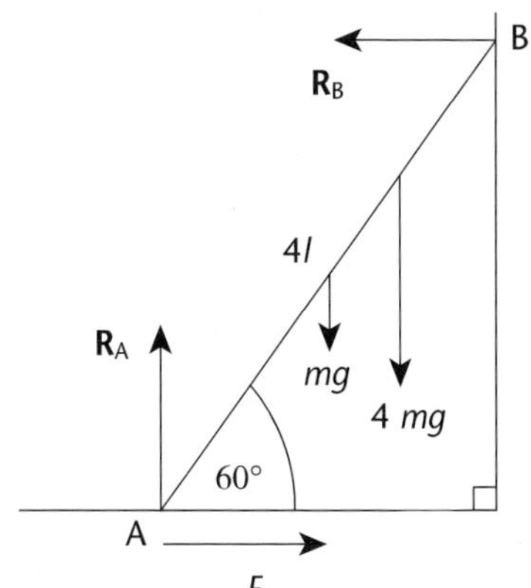

In this example, three equations may be formed by resolving the forces in two directions and taking moments.

The equations provide enough information to solve the problem.

Vertically: $\mathbf{R}_A = mg + 4mg = 5mg$ (1)

Horizontally: $F = \mathbf{R}_B$ (2)

M(A): $\mathbf{R}_B \times 4l \sin 60° = mg \times 2l \cos 60° + 4mg \times 3l \cos 60°$

giving: $\mathbf{R}_B \times 2\sqrt{3}l = mgl + 6mgl$

so: $\mathbf{R}_B = \dfrac{7mg}{2\sqrt{3}}$ (3)

From (2) and (3) $F = \dfrac{7mg}{2\sqrt{3}}$

Since the ladder is on the point of slipping, friction is limiting and the coefficient of friction is given by:

$$\mu = \frac{F}{\mathbf{R}_A} = \frac{7}{10\sqrt{3}} = \frac{7\sqrt{3}}{30}$$

It's a good idea to rationalise the denominator by multiplying the top and bottom of the fraction by $\sqrt{3}$.

Progress check

1 A uniform plank AB of length 3 m and mass 25 kg rests horizontally on two supports, one at each end. A mass of 15 kg is positioned on the plank 1 m from B. Find the reaction at each of the supports. Take $g = 9.8$ ms^{-2}.

2 A uniform ladder of length 5 m rests against a smooth wall with its base on rough horizontal ground. The coefficient of friction between the ladder and the ground is $\frac{2}{3}$ and the ladder is on the point of slipping. Show that the angle between the ladder and the ground is $\tan^{-1}(\frac{3}{4})$.

1 220.5 N (A), 171.5 N(B)

3.4 Centre of mass

After studying this section you should be able to:

- *find the centre of mass of a system of particles in one and two dimensions*
- *find the centre of mass of a lamina by considering an equivalent system of particles*
- *use centre of mass to determine conditions for the equilibrium of a lamina*

LEARNING SUMMARY

Centre of mass of a system of particles

AQA	M2
EDEXCEL	M2
OCR	M2
WJEC	M3, M5
NICCEA	M3

Key points from AS

- **Centre of mass**
  Revise AS page 83

If m_1 and m_2 are equal then this result gives the mid-point of the two masses.

In the diagram, m_1 and m_2 lie on a straight line through O. Their displacements from O are x_1 and x_2 respectively.

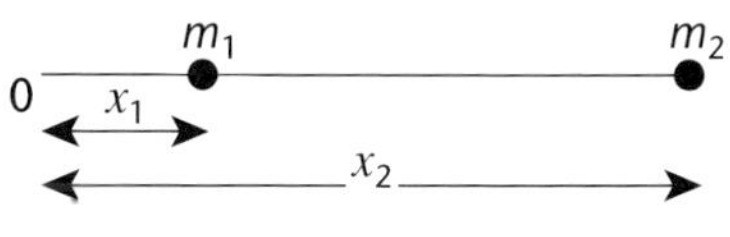

The position given by $\bar{x} = \dfrac{m_1x_1 + m_2x_2}{m_1 + m_2}$ is called the **centre of mass** of m_1 and m_2.

For n separate masses m_1, m_2, ... m_n the position of the centre of mass is given by

$$\bar{x} = \frac{m_1x_1 + m_2x_2 + \ldots m_nx_n}{m_1 + m_2 + \ldots m_n}$$ This is usually written as $$\bar{x} = \frac{\sum_{i=1}^{n} m_ix_i}{\sum_{i=1}^{n} m_i}$$

In two dimensions, using the usual Cartesian coordinates, the position of the centre of mass is at $(\bar{x}, \bar{y})$ where $\bar{x}$ and $\bar{y}$ are given by

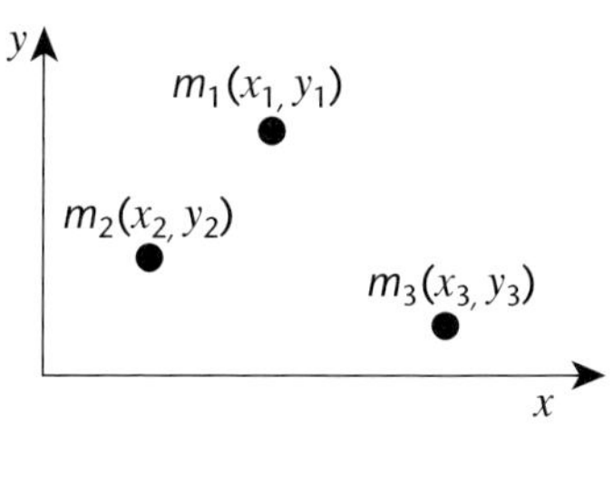

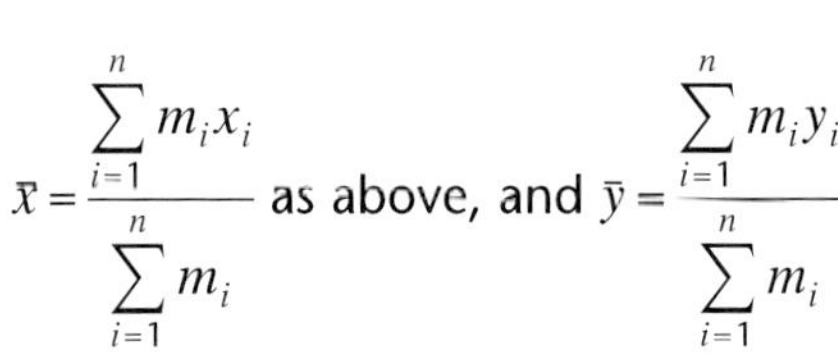

$$\bar{x} = \frac{\sum_{i=1}^{n} m_ix_i}{\sum_{i=1}^{n} m_i}$$ as above, and $$\bar{y} = \frac{\sum_{i=1}^{n} m_iy_i}{\sum_{i=1}^{n} m_i}$$

The method is easily extended to deal with any number of masses.

Example

Masses of 2 kg, 3 kg and 5 kg are positioned as shown in the diagram.
Find the coordinates of the centre of mass of the system.

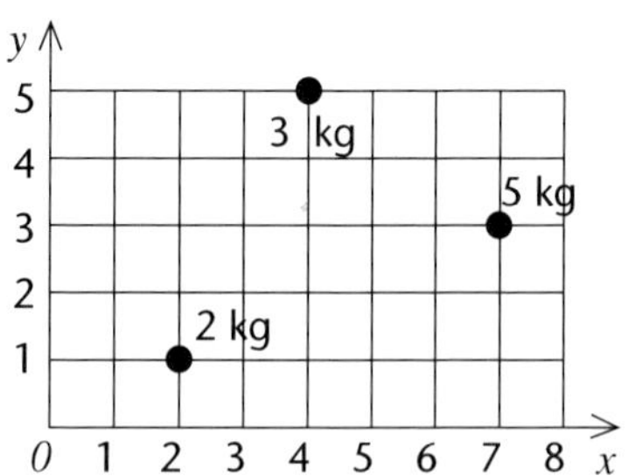

$$\bar{x} = \frac{2 \times 2 + 3 \times 4 + 5 \times 7}{2 + 3 + 5} = \frac{51}{10} = 5.1$$

$$\bar{y} = \frac{2 \times 1 + 3 \times 5 + 5 \times 3}{2 + 3 + 5} = \frac{32}{10} = 3.2$$

The centre of mass has coordinates (5.1, 3.2)

Centre of mass of a lamina

AQA	M2
EDEXCEL	M2
OCR	M2
WJEC	M3
NICCEA	M3

A **lamina** is something that is thin and flat such as a sheet of metal. The thickness of a lamina is ignored and it is treated as a two-dimensional object.

You can find the centre of mass of a uniform lamina by dividing it into parts which you then represent as particles. Each part is usually rectangular or triangular and in order to represent it as a particle:

- the mass of the particle is represented by the area of the part
- the position of the particle is taken to be at the geometric centre of the part.

This representation of a lamina gives you the information you need to apply the formulae for its centre of mass.

Example

Find the coordinates of the centre of mass of this uniform lamina.

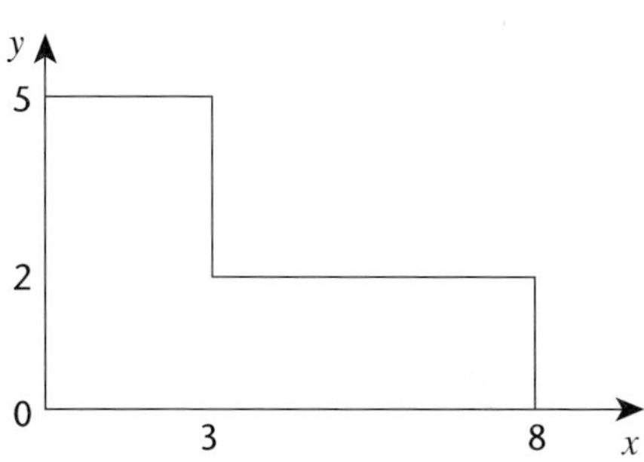

There are different ways to do this, but it doesn't matter which you choose.

This process is easily extended to deal with more complex composite figures.

The first step is to divide the lamina into rectangles.

Rectangle A has area 15 square units.
Its centre is at (1.5, 2.5).

Rectangle B has area 10 square units.
Its centre is at (5.5, 1).

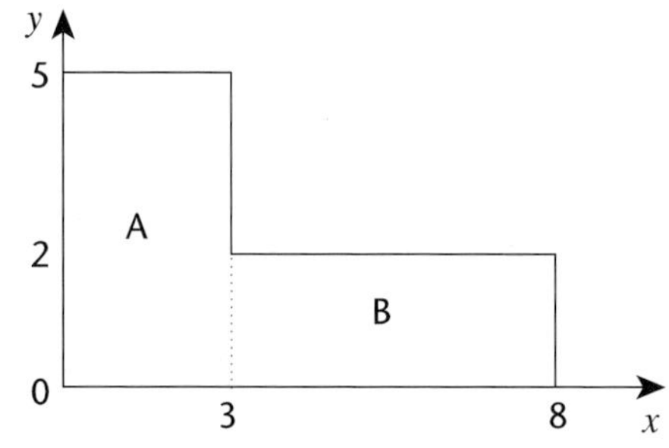

$$\bar{x} = \frac{15 \times 1.5 + 10 \times 5.5}{15 + 10} = 3.1$$

$$\bar{y} = \frac{15 \times 2.5 + 10 \times 1}{15 + 10} = 1.9$$

The centre of mass of the lamina has coordinates (3.1, 1.9).

Freely suspended lamina

AQA	M2
EDEXCEL	M2
OCR	M2
WJEC	M3
NICCEA	M3

Once you know where the centre of mass of a lamina is, you can use this to work out how it will move if it is freely suspended from a given point.

> When a lamina is freely suspended, it will move so that its centre of mass lies directly below the point of suspension.
>
> **KEY POINT**

Example

The diagram shows the lamina of the previous example hanging freely from one corner.

Find the size of angle θ between the direction of Oy and the horizontal.

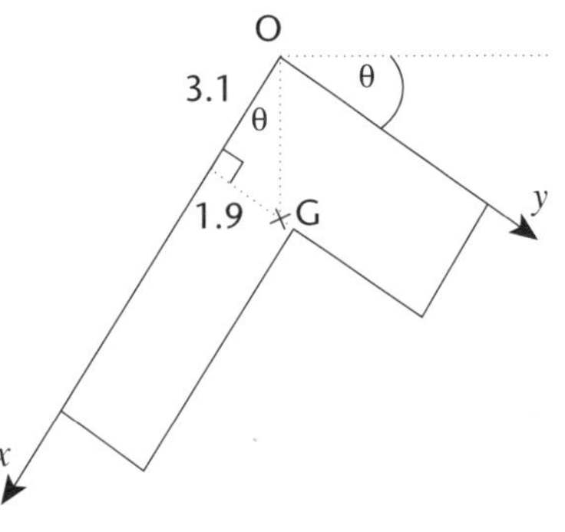

The position of the centre of mass is shown at the point G.
Notice that θ also lies in a right-angled triangle in which two of the sides are known.

From the diagram: $\tan \theta = \dfrac{1.9}{3.1}$

giving $\theta = 31.5°$ to 1 d.p.

Progress check

Masses of 2 kg, 3 kg and 5 kg are positioned as shown in the diagram.

(a) Find the coordinates of G, the centre of mass of the system.

(b) Find the angle θ between OG and Ox.

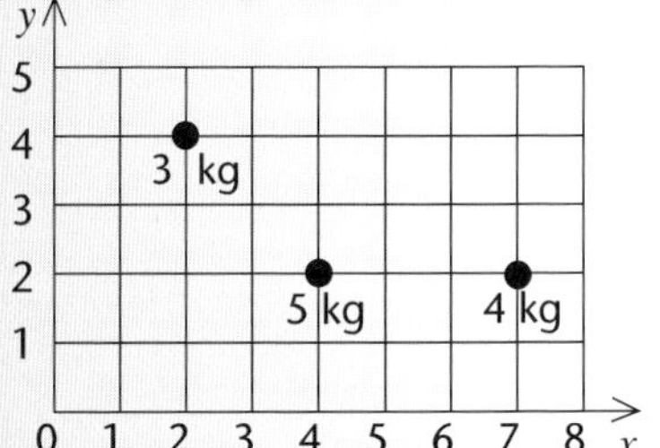

(a) G(4.5, 2.5) (b) 29.1°

3.5 Collisions and impulse

After studying this section you should be able to:

- *apply the principle of conservation of momentum to direct impact*
- *apply Newton's experimental law using the coefficient of restitution*
- *calculate the loss of mechanical energy in a collision*
- *recall and use the definition of impulse as a change of momentum*

LEARNING SUMMARY

Conservation of momentum

EDEXCEL	M2
OCR	M2
WJEC	M2
NICCEA	M3

The **momentum** of an object is the product of its mass and its velocity, i.e. momentum = $m\mathbf{v}$. Momentum has direction and so it is a vector quantity.

Momentum is measured in Ns (Newton seconds).

It's important to use the correct units. Remember that speed is the magnitude of the velocity.

For example, an object of mass 5 kg moving with speed 6 ms^{-1} in a given direction has momentum $5 \times 6 = 30$ Ns in the direction of movement.

The **Principle of conservation of momentum** states that, in a collision, the total momentum before impact = the total momentum after impact.
Remember:

Key points from AS

- **Momentum**
  Revise AS page 86

- always draw a diagram to represent the situation
- choose one direction to be positive (usually left to right $\longrightarrow$)
- show any unknown velocities in the positive direction.

Example

When particles coalesce, they stick together and behave as a single particle.

Two particles A and B of masses 4 kg and 6 kg respectively move towards each other along the same straight line. Particle A is moving with speed 3 ms^{-1} and particle B is moving with speed 5 ms^{-1}. Given that the particles coalesce on impact, find their common velocity immediately afterwards.

The velocity of B is in the negative direction.

Before impact:

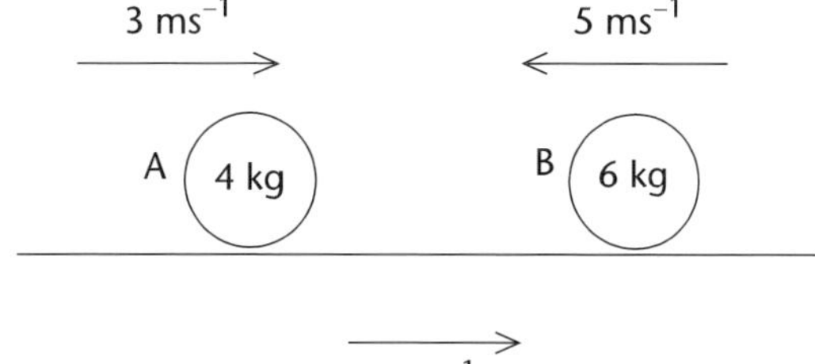

After impact:

v ms^{-1} $\longrightarrow$

By the principle of conservation of momentum:

$$4 \times 3 - 6 \times 5 = 10v$$

giving: $10v = -18$

so: $v = -1.8$

v is negative so the direction of the velocity is from right to left.

The particles move with velocity 1.8 ms^{-1} in the direction from B to A.

Newton's experimental law

EDEXCEL M2
OCR M2
NICCEA M3

Some problems require two unknown values to be found and in this situation a second equation is needed.

Newton found by experiment that, when two objects collide, the ratio of their speed of separation to their speed of approach has a fixed value. The symbol used to represent this ratio is e and its value is depends on what the objects are made of.

$$e = \frac{\text{speed of separation}}{\text{speed of approach}}$$

This is known as **Newton's experimental law** and is often used to provide a second equation.

This is often used in the form $e \times$ speed of approach = speed of separation.

The speed of separation is always less than or equal to the speed of approach so it follows that $0 \leqslant e \leqslant 1$. The ratio e is called the **coefficient of restitution.**

Example

In the exam, you may need to deal with more than one impact. See the sample exam questions p. 97.

A smooth sphere P moves towards a stationary smooth sphere Q in a straight line through their line of centres. P has mass 2 kg and speed 5 ms^{-1}. Q has mass 4 kg. Given that the coefficient of restitution between the spheres is 0.5, find:

(a) the speed of each sphere immediately after impact
(b) the loss of energy in the collision.

(a)

Before impact: P 2 kg moving at 5 ms^{-1} →; Q 4 kg

After impact: v_1 →; v_2 →

Momentum is conserved: $2 \times 5 + 4 \times 0 = 2v_1 + 4v_2$

So: $v_1 + 2v_2 = 5$ (1)

Newton's law: $0.5 \times 5 = v_2 - v_1$ (2)

Adding (1) and (2) gives: $3v_2 = 7.5 \Rightarrow v_2 = 2.5$

and: $v_1 = 0$

Immediately after the impact, P is brought to rest and Q moves with speed 2.5 ms^{-1}.

Kinetic energy = $\frac{1}{2}mv^2$. Even though momentum is conserved, energy is lost.

(b) Loss of energy = KE before impact – KE after impact
$= 0.5 \times 2 \times 5^2 - 0.5 \times 4 \times 2.5^2$
$= 12.5$ J

Impulse

EDEXCEL M2
OCR M2
WJEC M2
NICCEA M3

Key points from AS

- **Impulse**
  Revise AS page 86

Impulse = change in momentum

When two particles collide, each receives an impulse from the other of equal size but opposite sign. In this way, the total change in momentum is zero. We would expect this to happen because the total momentum of the system is conserved in a collision.

Example

A ball of mass 0.5 kg moving with velocity $6\mathbf{i}+2\mathbf{j}$ ms^{-1} is kicked and receives an impulse of $2\mathbf{i}+11\mathbf{j}$ Ns. Find the velocity of the ball immediately after it is kicked.

Momentum immediately after kick = initial momentum + impulse

$= 0.5(6\mathbf{i}+2\mathbf{j})+2\mathbf{i}+11\mathbf{j}=5\mathbf{i}+12\mathbf{j}$

$= 0.5\mathbf{v}$ (where $\mathbf{v}$ is the velocity)

This gives: $\mathbf{v}=10\mathbf{i}+24\mathbf{j}$ ms^{-1}

Progress check

Two particles A and B move towards each other with speeds of 6 ms^{-1} and 2 ms^{-1} respectively. A has mass 3 kg and B has mass 1 kg. The coefficient of restitution between the particles is 0.6. Find the speed of each particle after the collision.

A 2.8 ms^{-1}, B 7.6 ms^{-1}.

3.6 Uniform circular motion

After studying this section you should be able to:

- understand angular speed and be able to apply the formula $v = r\omega$
- understand that a particle moving in a circular path with constant speed has an acceleration towards the centre of the circle
- use the formula $\omega^2 r$, or its equivalent form $\frac{v^2}{r}$, to represent the magnitude of the acceleration towards the centre of the circle in solving problems

LEARNING SUMMARY

Angular speed

AQA	M3
EDEXCEL	M3
OCR	M2
WJEC	M3
NICCEA	M2

Suppose that a particle P moves with constant speed in a circular path with centre O.

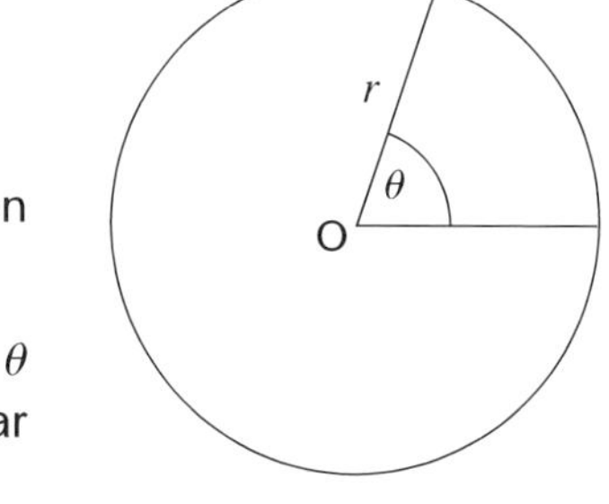

The direction of OP changes with time.
The angle θ, through which OP turns in some given time, is usually measured in radians.

The **angular speed** of P is the rate of change of θ with respect to time. Using ω to represent angular speed in radians per second gives:

$$\omega = \frac{\theta}{t}$$

One advantage of using radians is that the relationship between angular speed ω, and linear speed v, is easily expressed as $v = rw$.

For example, a particle moving in a circular path of radius 2 m with angular speed 10 radians/sec has a straight line speed of $2 \times 10 = 20$ ms^{-1}.

Radial acceleration

AQA	M3
EDEXCEL	M3
OCR	M2
WJEC	M3
NICCEA	M2

Notice that while the speed of the particle may be constant, its *direction* changes as it moves around the circle. This means that the *velocity* is *not constant* and so the particle must be *accelerating*.

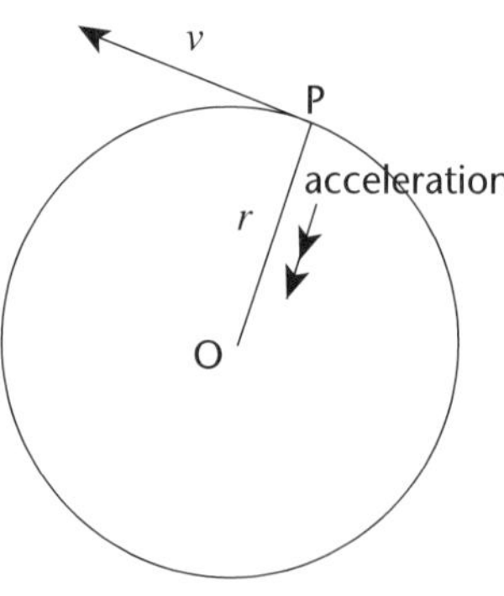

This acceleration is always directed towards the centre of the circle and is known as **radial acceleration** or sometimes **centripetal acceleration**.

The magnitude of this acceleration is given by $a = \omega^2 r$.

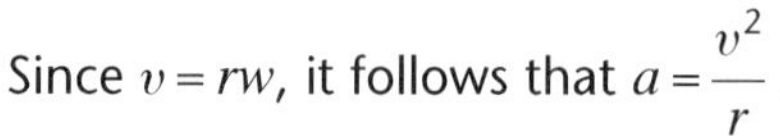
Since $v = rw$, it follows that $a = \frac{v^2}{r}$.

This means that there is a choice of which formula to use. In practice, you should use the one that best fits the available information.

Example

Since we are given the angular speed, the formula $a = \omega^2 r$ is the simpler one to use here.

A particle moves in a circular path of radius 3 m with angular speed 5 radians/sec. Find the magnitude of its radial acceleration.

Using $a = \omega^2 r$: the radial acceleration is $5^2 \times 3 = 75$ ms^{-2}.

Circular motion and force

AQA	M3
EDEXCEL	M3
OCR	M2
WJEC	M3
NICCEA	M2

The radial acceleration of an object moving in a circular path must be produced as the result of a *force acting towards the centre of the circle*. This force is sometimes called the **centripetal force**.

Using $F = ma$, the force needed to keep an object of mass m kg moving in a circular path is $m\omega^2 r$. An alternative form of this is $\frac{mv^2}{r}$.

In applying this theory to solving problems, the centripetal force may take various forms such as a tension in a string, friction or the gravitational attraction of a planet.

Example

One end of a string of length 80 cm is attached to a point O on a smooth horizontal table. The other end is attached to a mass of 3 kg moving with speed 4 ms^{-1} and the string is taut. Find the tension in the string.

It is important to write the length of the string in metres.

Using $T = \frac{mv^2}{r}$ gives $T = \frac{3 \times 4^2}{0.8} = 60$

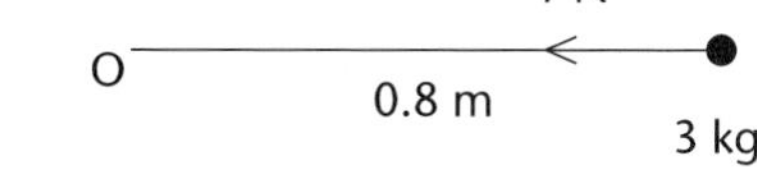

The tension in the string is 60 N.

The conical pendulum

AQA	M3
EDEXCEL	M3
OCR	M2
WJEC	M3
NICCEA	M2

One particular situation involving circular motion is described as the **conical pendulum**.

This is where an object moves in a horizontal circle while suspended by a string attached to a point vertically above the centre of the circle.

- The vertical component of the tension in the string supports the weight of the object.
- The horizontal component of the tension in the string provides the centripetal force.

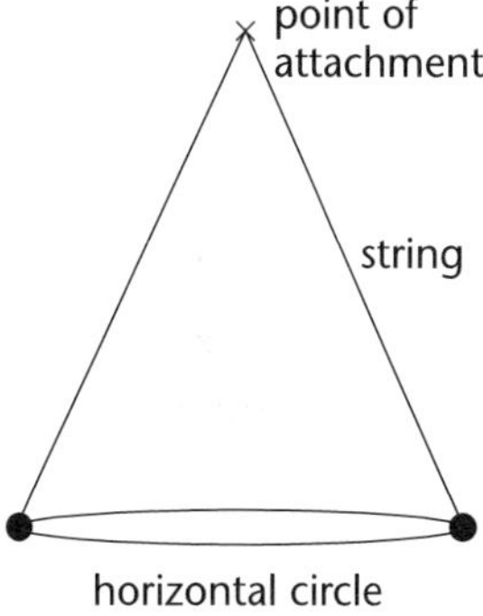

Example

A light inextensible string is fixed at one end to a point O. The other end is attached to a particle of mass 2 kg which moves in a horizontal circle of radius 30 cm with angular speed 4 radians per second. Find the angle of inclination of the string to the vertical. Take $g = 9.8$ ms^{-2}.

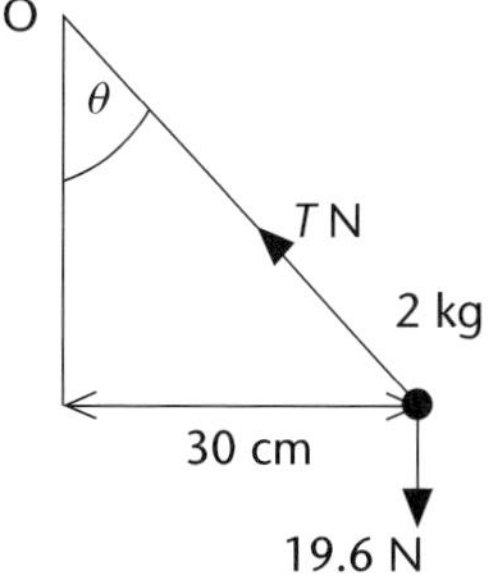

Vertically: $T \cos\theta = 19.6$ (1)

Horizontally: $T \sin\theta = 2 \times 4^2 \times 0.3 = 9.6$ (2)

(2) ÷ (1) gives: $\tan\theta = \frac{9.6}{19.6} \Rightarrow \theta = 26.1°$

The string is inclined at 26.1° to the vertical.

Progress check

One end of a string of length 60 cm is attached to a point O on a smooth horizontal table. The other end is attached to a mass of 5 kg moving with speed 3 ms^{-1} and the string is taut. Find the tension in the string.

1 75 N

3.7 Work, energy and power

After studying this section you should be able to:

- *understand kinetic and potential energy and the work energy principle*
- *understand the principle of conservation of mechanical energy*
- *understand the definition of power*
- *solve problems involving work energy and power*

LEARNING SUMMARY

Work and energy

AQA	M2
EDEXCEL	M2
OCR	M2
WJEC	M2, M5
NICCEA	M2

In everyday language, **energy** is needed to do **work**. This idea is given a formal interpretation within mechanics.

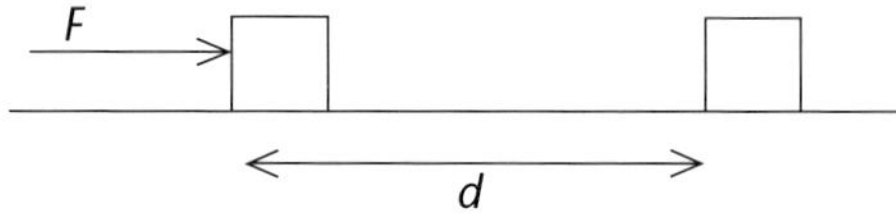

The situation where the force is not constant is dealt with in Mechanics 3 of this book.

The work done, W, by a *constant* force, F, in moving an object through a distance, d, along its line of action is given by $W = Fd$. The unit of work is the **joule** (J).

Energy is also measured in joules and the energy used in moving the object is equal to the work done. In a way, work and energy are like two sides of the same coin.

Energy may take different forms such as heat, sound, electrical, chemical and mechanical. An important property of energy is that it may change from one form to another. For your mechanics module, the focus of attention is on mechanical energy.

There are two forms of mechanical energy: **kinetic energy (KE)** and **potential energy (PE)**

The kinetic energy of an object is the energy it has due to its *motion*.

Its value depends on the mass of the object and its speed. KE = $\frac{1}{2}mv^2$.

For example, a mass of 4 kg moving with speed 5 ms^{-1} has KE = $\frac{1}{2} \times 4 \times 5^2 = 50$ J

The potential energy of an object is the energy it has due to its *position*.

Potential energy may be gravitational or elastic. Elastic potential energy is covered in Mechanics 3 in this book.

The gravitational potential energy of an object, relative to some level, is equal to the work done against gravity in raising the object from that level to its current position.

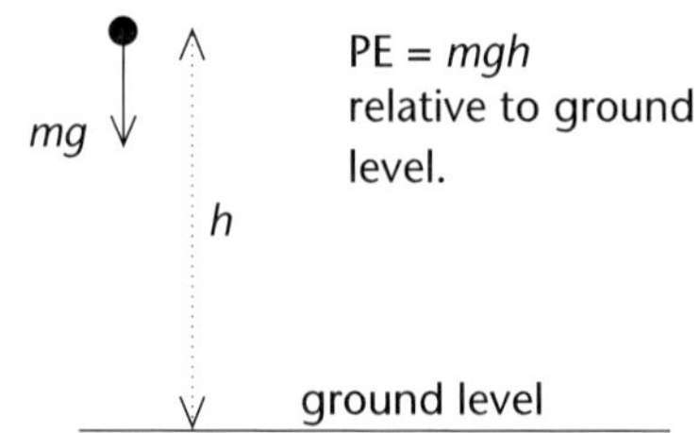

> The total mechanical energy of an object is the sum of its KE and PE.
>
> The **principle of conservation of mechanical energy** states that the total mechanical energy of a system remains constant, provided that it is not acted on by any external force, other than gravity.
>
> KEY POINT

The principle of conservation of energy applies, for example, to the motion of a projectile. Any change in the PE of the projectile corresponds to an equal but opposite change in its KE.

Power

AQA	M2
EDEXCEL	M2
OCR	M2
WJEC	M2
NICCEA	M2

Power is the rate of doing work. It is measured in watts (W) where 1 W = 1 Js^{-1}.

If the point of application of a force F moves with speed v in the *direction of the force* then the power P is given by $P = Fv$.

Example

Find the power generated by a car engine when the car travels at a constant speed of 72 kmh^{-1} on horizontal ground against resistance forces of 2500 N.

Since the speed is constant, the force produced by the engine must match the resistance force.

$$72 \text{ kmh}^{-1} = \frac{72 \times 1000}{60 \times 60} \text{ ms}^{-1} = 20 \text{ ms}^{-1}.$$

$$\begin{aligned} \text{Power} &= 2500 \times 20 \text{ W} = 50\,000 \text{ W} \\ &= 50 \text{ kW} \end{aligned}$$

Exam questions may involve resisted motion on an inclined plane.

Example

A car of mass 1000 kg travels up a hill inclined at angle θ to the horizontal, where $\sin\theta = \frac{1}{20}$. The non-gravitational resistance to motion is 2000 N and the power output from the engine is 60 kW. Find the acceleration of the car when it is travelling at 10 ms^{-1}. Take $g = 10 \text{ ms}^{-2}$.

A clearly labelled diagram is a must.

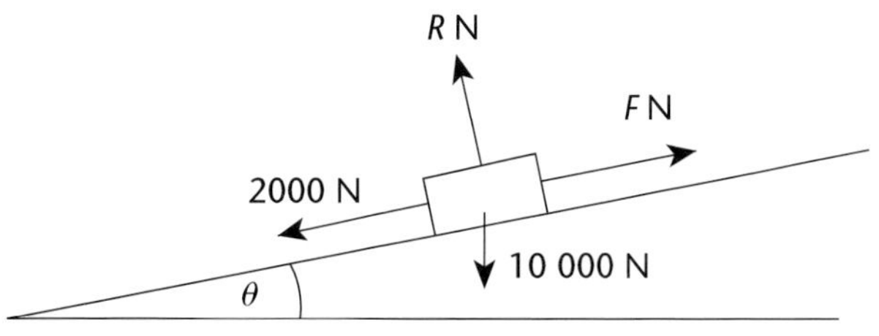

In this formula, F represents the *resultant force* up the hill.

Using $P = Fv$: $60\,000 = F \times 10 \Rightarrow F = 6000$

Using $F = ma$: $6000 - 2000 - 10\,000 \times \frac{1}{20} = 1000a$

This gives: $3500 = 1000a \Rightarrow a = 3.5$

The acceleration of the car up the hill is 3.5 ms^{-2}.

Progress check

1 A ball is kicked from ground level with speed 15 ms^{-1} and just clears a wall of height 2 m. The speed of the ball as it passes over the wall is v ms^{-1}.

(a) Assume that the ball has mass m kg and use the conservation of mechanical energy principle to write an equation.

(b) Solve the equation to find v.

2 A car of mass 800 kg travels up a hill inclined at angle θ to the horizontal, where $\sin\theta = \frac{1}{20}$. The non-gravitational resistance to motion is 1600 N and the power output from the engine is 45 kW. Find the acceleration of the car when it is travelling at 15 ms^{-1}. Take $g = 10 \text{ ms}^{-2}$.

1 (a) $\frac{1}{2}m \times 15^2 = m \times 9.8 \times 2 + \frac{1}{2}mv^2$

(b) 13.6 ms^{-1}.

2 1.25 ms^{-2}.

Sample questions and model answers

1

A ball of mass 250 g is released from rest 2 m above the ground. The coefficient of restitution between the ball and the ground is 0.7. Take $g = 9.8 \text{ ms}^{-2}$.

(a) Find the height reached by the ball as it rebounds from the ground.

(b) Find the impulse exerted on the ball by the ground.

(a) Using $v^2 = u^2 + 2as$

gives $v^2 = 2 \times 9.8 \times 2$

$= 39.2$

so $v = \sqrt{39.2}$

There is no need to work out $\sqrt{39.2}$ at this stage.

The ball strikes the floor with speed $\sqrt{39.2}$ ms^{-1}.

Using Newton's law, the ball rebounds with speed $\sqrt{39.2} \times 0.7$ ms^{-1}

Some simple statements help to make your method clear.

Using $v^2 = u^2 + 2as$

gives $0 = 39.2 \times 0.49 - 2 \times 9.8 \times h$

so $h = \dfrac{39.2 \times 0.49}{2 \times 9.8} = 0.98$

The ball reaches a height of 0.98 m when it rebounds from the ground.

(b) Taking upwards as positive:

Momentum of ball immediately before it hits the ground

$= -0.25 \times \sqrt{39.2}$ Ns

Write the mass in kg

Momentum of ball immediately after it hits the ground

$= 0.25 \times \sqrt{39.2} \times 0.7$ Ns

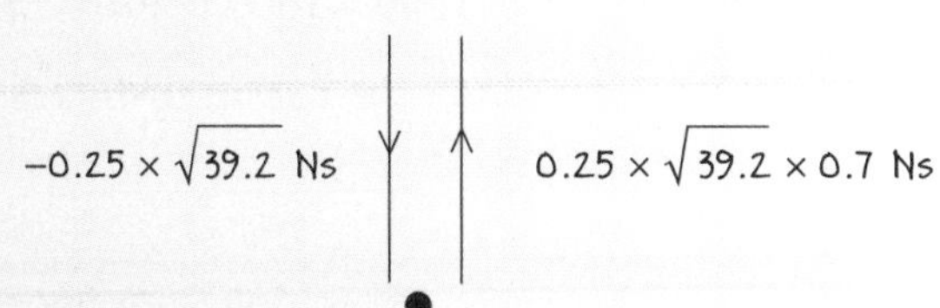

Impulse = change in momentum

$= 0.25 \times \sqrt{39.2} \times 0.7 - (-0.25 \times \sqrt{39.2})$

$= 2.66$ Ns to 3 s.f.

The impulse exerted on the ball by the ground is 2.66 Ns.

Sample questions and model answers (continued)

2

A particle P of mass 3 m is moving with speed $2u$ when it strikes a particle Q of mass m which is at rest. The coefficient of restitution between the particles is e.

(a) Show that the speed of P after the collision is $\frac{v}{2}(3-e)$ and find the speed of Q.

(b) Q subsequently strikes a wall perpendicular to its direction of motion and rebounds to so that a second collision with P occurs. Given that the coefficient of restitution between Q and the wall is $\frac{1}{3}$ find the speed of P after the second collision.

(c) Show that P moves in the same direction after the second collision.

(a) Before impact: P (3m) moving with speed $2u$ →; Q (m) at rest.

After impact: P with v_P →; Q with v_Q →.

The approach is exactly the same as when numerical values of the speeds are given.

Conservation of momentum gives: $6mu = 3mv_P + mv_Q$

which simplifies to $6u = 3v_P + v_Q$ (1)

Newton's experimental law gives: $2ue = v_Q - v_P$ (2)

(1) – (2) gives: $6u - 2ue = 4v_P$

This is the result given in the question.

giving $v_P = \frac{u}{2}(3-e)$ as required

and from (2) $v_Q = 2ue + v_P = 2ue + \frac{u}{2}(3-e)$

$= \frac{u}{2}(4e+3-e) = \frac{3u}{2}(e+1)$

The speed of Q after the collision is $\frac{3u}{2}(e+1)$.

(b) The speed of Q after hitting the wall is $\frac{1}{3} \times \frac{3u}{2}(e+1) = \frac{u}{2}(e+1)$ in the opposite direction.

Repeat the process with the new speeds. Draw a diagram and set your working out clearly.

Before impact: P (3m) moving with speed $\frac{u}{2}(3-e)$ →; Q (m) moving with speed $\frac{u}{2}(e+1)$ ←.

After impact: P with w_P →; Q with w_Q →.

Conservation of momentum gives: $\frac{3mu}{2}(3-e) - \frac{mu}{2}(e+1) = 3mw_P + mw_Q$

which simplifies to $4u - 2ue = 3\,w_P + w_Q$ (3)

Newton's experimental law gives: $\left(\frac{u}{2}(3-e) + \frac{u}{2}(e+1)\right)e = w_Q - w_P$

which simplifies to $2u = w_Q - w_P$ (4)

(3) – (4) gives $2u - 2ue = 4\,w_P$

and $w_P = \frac{u}{2}(1-e)$

There is no need to find w_Q.

so the speed of P immediately after the second collision is $w_P = \frac{u}{2}(1-e)$.

(c) $1 - e \geqslant 0$ since $0 \leqslant e \leqslant 1$ so the direction of movement of P is not changed.

Practice examination questions

1 A ball is struck with speed 25 ms^{-1} from a point 50 cm above the ground. When it has travelled 20 m horizontally, the ball just clears a fence that is 3 m high.

(a) Show that a possible angle of projection of the ball is 16.5°.

(b) Find the distance d m, between the fence and the point where the ball strikes the ground for this angle of projection.

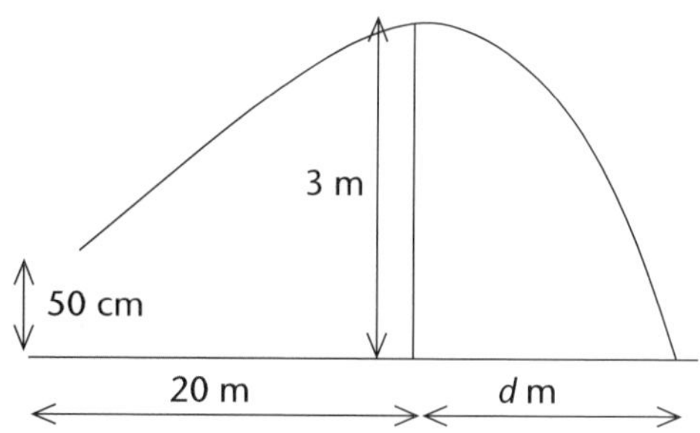

2 A particle moves along the x axis such that at time t seconds its displacement from a point O on the line is $x = t^3 - 6t^2 + 3$ metres.

(a) Show that the particle is initially at rest and find its greatest distance from O in the negative direction.

(b) Find the speed of P when it passes through its initial position.

3 The position vector of a particle P at time t seconds relative to a fixed origin O is given by $\mathbf{r} = (4t^2 + 5)\mathbf{i} + 2t^3\mathbf{j}$ metres.

Find:

(a) the speed of P when $t = 1$

(b) the time when P is moving parallel to the vector $\mathbf{i} + 3\mathbf{j}$.

4 The diagram shows a uniform lamina ABCDEFGH.

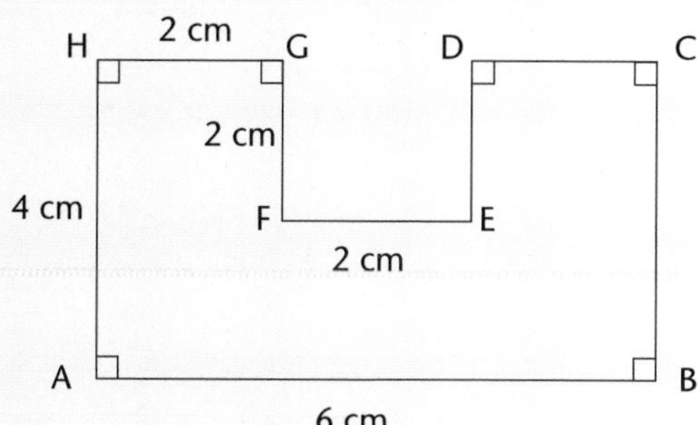

(a) Write down the distance of the centre of mass of the lamina from AH.

(b) Find the distance of the centre of mass of the lamina from AB.

(c) The lamina is now suspended freely from A. Find the angle that AB makes with the horizontal in its equilibrium position.

Practice examination questions (continued)

5

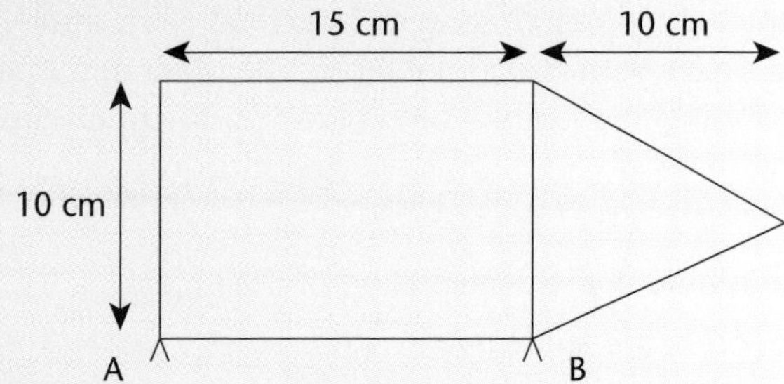

The diagram shows a uniform lamina of mass 0.8 kg resting on supports at A and B.

(a) Find the horizontal distance of the centre of mass from the support at A.

(b) Find the magnitude of the reaction force on the lamina at each support.

6

A and B are two smooth spheres of equal size and they are approaching each other on a direct line through their centres. A is moving with speed 4 ms^{-1} and B is moving with speed 2 ms^{-1}. The coefficient of restitution between the spheres is $\frac{1}{3}$.

Find:

(a) the speed of each sphere after the collision

(b) the impulse exerted by A on B during the collision

(c) the loss of mechanical energy during the collision.

7 Two particles P and Q approach each other with speeds u and $2u$ respectively. P has mass $3m$ and Q has mass m. The coefficient of restitution between the particles is e.

Given that the direction of motion of P is reversed in the collision, show that $e > \frac{1}{3}$.

8 A motorbike is driven round a corner in a circular path of radius 40 m. The maximum speed that this can be done without skidding is 15 ms^{-1}.

Calculate the coefficient of friction between the tyres and the road surface to 2 s.f.

Take $g = 9.8$ ms^{-2}.

Practice examination questions (continued)

9 One end of a light inelastic string of length 50 cm is attached to a fixed point 40 cm above a smooth horizontal table. The other end is attached to an object of mass 2 kg moving in a circular path of radius 30 cm on the table.

Given that the angular speed of the object is 3 radians per second, find:

(a) the magnitude of the acceleration of the object

(b) the tension in the string

(c) the reaction force of the table on the object.

10 A particle P of mass m is attached to one end of a light inelastic string of length l. The other end of the string is attached to a fixed point O. P is released from rest with the string taut and horizontal.

(a) Explain why, in the subsequent motion, the tension in the string does no work on the particle.

(b) Explain why it is not valid to use the formula $v^2 = u^2 + 2as$ in this situation.

(c) Find the maximum speed of the particle assuming that there is no resistance to motion.

11 A car of mass 900 kg travels up a straight road inclined at angle α to the horizontal, where $\sin \alpha = \frac{1}{15}$. Assume that the total non-gravitational force opposing the car's motion has a constant value of 800 N. Take $g = 9.8$ ms^{-2}.

(a) Find the driving force required for the car to maintain a steady speed up the slope.

(b) Find the power output of the car's engine given that its speed up the slope has a constant value of 20 ms^{-1}.

(c) At the top of the hill, the road surface is horizontal. If the power output of the engine remains the same, what is the initial acceleration of the car?

12 A football of mass 0.2 kg is travelling horizontally towards the goal with speed 20 ms^{-1} when it is struck by the keeper. It then rebounds vertically with speed 5 ms^{-1}.

Find the magnitude and direction of the impulse given to the ball by the keeper.

Chapter 4

Mechanics 3

The following topics are covered in this chapter:

- *Linear motion with variable acceleration*
- *Elastic springs and strings*
- *Simple harmonic motion*
- *Further circular motion*
- *Centre of mass of uniform solids*
- *Collisions in two dimensions*

4.1 Linear motion with variable acceleration

After studying this section you should be able to:

- *set up and solve a differential equation representing acceleration in a straight line, where the acceleration is given as a function of displacement or time*
- *solve problems which can be modelled as the linear motion of a particle under the action of a variable force*

LEARNING SUMMARY

Kinematics

AQA	M3
EDEXCEL	M3
OCR	M3
WJEC	M2, M3
NICCEA	M2

Kinematics is the branch of mathematics that deals with the motion of a particle in terms of displacement, velocity and acceleration, without considering the forces that may be required to cause the motion.

For motion in a straight line, the distinction between **distance** and **displacement** is only that displacement may be positive or negative to indicate a sense of direction, whereas distance is always taken to be positive.

The same distinction applies to **speed** and **velocity**. Velocity in a straight line may be positive or negative, depending on the direction of movement, but speed is always taken to be positive.

In the standard notation:

- displacement is represented by x
- velocity is represented by $v = \dot{x} = \dfrac{dx}{dt}$.

You may need to set up and solve equations of the form $\dfrac{dx}{dt} = f(x)$ or $\dfrac{dx}{dt} = f(t)$.

- Acceleration is represented by $\ddot{x} = \dfrac{dv}{dt} = v\dfrac{dv}{dx}$.

This is an important result and you may need to establish it in the exam.

You should recognise that by the chain rule for differentiation: $\dfrac{dv}{dt} = \dfrac{dv}{dx} \times \dfrac{dx}{dt} = v\dfrac{dv}{dx}$.

You may need to set up and solve equations of the form $\dfrac{dv}{dt} = f(t)$ or $v\dfrac{dv}{dx} = f(x)$.

This requires a differential equation because the acceleration is not constant.

Example

A particle P moves in the direction of the positive x-axis with acceleration $3\sqrt{x}$ ms^{-2} directed away from O, where OP = x metres. When $x = 1$ the speed of P is 2 ms^{-1}. Find its speed when $x = 4$.

See **separation of variables** p. 38.

The acceleration is $v\dfrac{dv}{dx} = 3x^{\frac{1}{2}} \Rightarrow \int v\,dv = \int 3x^{\frac{1}{2}}\,dx$

This gives: $\frac{v^2}{2} = 2x^{\frac{3}{2}} + c$

When $x = 1$, $v = 2$ so $2 = 2 + c \Rightarrow c = 0$.

so: $v^2 = 4x^{\frac{3}{2}}$

When $x = 4$, $v^2 = 32 \Rightarrow v = 4\sqrt{2}$

The particle has speed $4\sqrt{2}$ ms^{-1} when $x = 4$.

Dynamics

AQA	M3
EDEXCEL	M3
OCR	M3
WJEC	M2, M3
NICCEA	M2

Dynamics is concerned with the motion of a particle in response to forces that act upon it. The first step is to identify all of the forces and then use $F = ma$ to set up a differential equation.

Example

A particle P of mass 0.2 kg is attached to one end of a light elastic string. The other end of the string is attached to a fixed point on a smooth horizontal table. P is released from rest and moves along the surface of the table towards a point O.

The tension in the string has magnitude $25x$ N where x metres is the displacement of P from O. Given that $x = 2$ at the point of release, find the speed of P when $x = 1$.

Draw a diagram and label the forces.

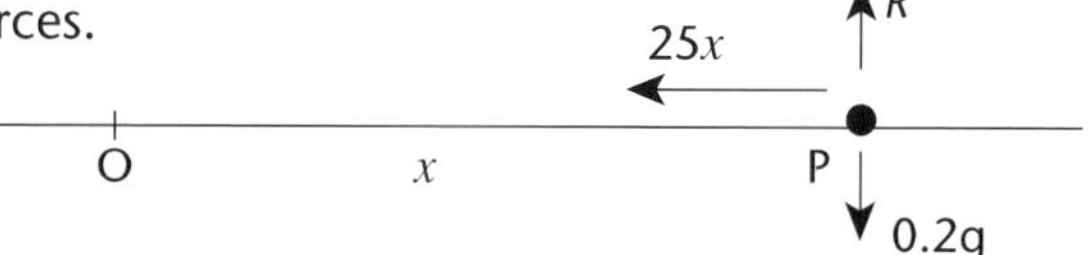

In the diagram, x increases from left to right so take this as the positive direction.

Using $F = ma$: $-25x = 0.2v\frac{dv}{dx}$

so: $\int -125x\,dx = \int v\,dv$

giving: $\frac{v^2}{2} = -125\frac{x^2}{2} + c$

Simplifying the equation can make it easier to substitute particular values.

This may be written as $v^2 = A - 125x^2$

When $x = 2, v = 0$ so $0 = A - 125 \times 4 \Rightarrow A = 500$

When $x = 1$, $v^2 = 500 - 125 = 375$

giving: $v = \sqrt{375}$

The speed of P when $x = 1$ is 19.4 ms^{-1} to 3 s.f.

Progress check

1 A particle P of mass 0.4 kg moves in the direction of the positive x-axis with acceleration $\frac{5}{x}$ ms^{-2} away from O, where OP = x metres. When $x = 1$ the speed of P is 3 ms^{-1}. Find its speed when $x = 5$.

2 A particle P of mass 0.5 kg moves in a straight line, away from a point O, under the action of a force of magnitude $\frac{x^2}{12}$ N directed towards O, where OP = x metres. When $x = 3$ the speed of P is 15 ms^{-1}. Find x when P is brought to rest.

1 20.3 ms^{-1} to 3 s.f.
2 12.7 m to 3 s.f.

4.2 Elastic springs and strings

After studying this section you should be able to:

- *understand and apply Hooke's law, including the term modulus of elasticity*
- *find the energy stored in an elastic string or spring*
- *solve problems involving elastic strings and springs using the work-energy principle*

LEARNING SUMMARY

Hooke's law

AQA	M3
EDEXCEL	M3
OCR	M3
WJEC	M3
NICCEA	M2

An elastic string will vary in length depending on its tension. Its **natural length** l, is its length when the tension is zero. The *extra* length of an elastic string due to a tension T within it is called the **extension** x.

Hooke's law states that the tension in an elastic string is proportional to the extension. This may be expressed as $T \propto x$ or

$$T = kx \qquad (1)$$

for some constant value k.

When a string has doubled its length, the extension must equal the natural length.

The **modulus of elasticity** λ, is the tension required to make an elastic string *double its length*. It follows from (1) that $\lambda = kl$ and $k = \frac{\lambda}{l}$.

Hooke's law is normally used in this form.

Substituting for k in (1): Hooke's law may now be written as $T = \frac{\lambda}{l}x$.

Hooke's law applies in the same way to **springs** that are under tension.
The difference is that a spring will also exert an outward force, or thrust, when it is *compressed* by a distance x. The formula given by Hooke's law applies in both situations.

Example

A light elastic string of natural length 0.8 m and modulus of elasticity 30 N is extended to a length of 1.1 m. Find the tension in the string.

In this case: $l = 0.8$, $\lambda = 30$ and $x = 0.3$.

Using Hooke's law gives: $T = \frac{30}{0.8} \times 0.3 = 11.25$

The tension in the string is 11.25 N.

Example

A spring of natural length 0.5 m is compressed to a length of 0.35 m by a force of 60 N. Find the modulus of elasticity.

For the spring: $l = 0.5$, $x = 0.15$ and $T = 60$.

Using Hooke's law gives: $60 = \frac{\lambda}{0.5} \times 0.15 \Rightarrow 0.3\lambda = 60$

so: $\lambda = 200$

The modulus of elasticity for the spring is 200 N.

Elastic potential energy

AQA	M3
EDEXCEL	M3
OCR	M3
WJEC	M3
NICCEA	M2

The elastic potential energy EPE of an elastic string or spring is its capacity for doing work as it returns to its natural length. Assuming that there is no loss of mechanical energy, this is equivalent to the work done in producing the extension or compression in the first place.

The area under the graph represents the work done against the tension in producing a given extension.

The diagram shows the linear relationship between T and x.

Work done $= \frac{1}{2} \times x \times \frac{\lambda}{l} x$

so the EPE $= \frac{\lambda x^2}{2l}$

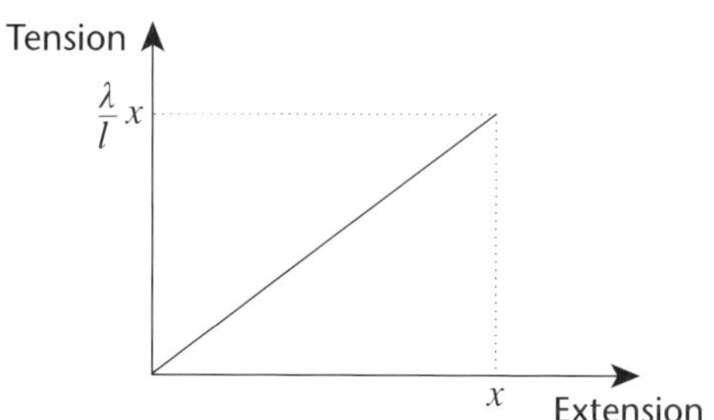

Example

An elastic string of natural length 1 m and modulus of elasticity 30 N is stretched to a length of 1.4 m. Calculate the EPE stored in the string.

EPE is measured in Joules.

$$\text{EPE} = \frac{\lambda x^2}{2l} = \frac{30 \times 0.4^2}{2 \times 1} = 2.4 \text{ J.}$$

Example

One end of a light elastic string is attached to a point O, on a smooth horizontal table. The other end is attached to a particle P of mass 0.3 kg held in position on the table such that OP = 1.8 m. The string has modulus of elasticity 20 N and natural length 1.2 m.

The system is released from rest. Find the speed of P when the string becomes slack.

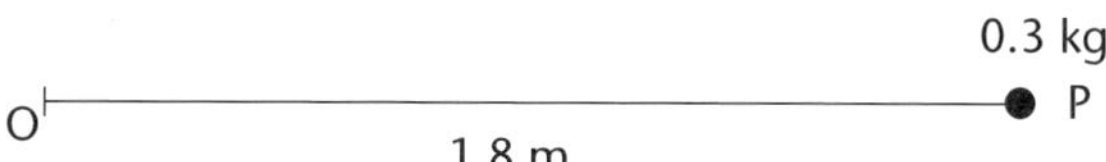

$$\text{EPE} = \frac{\lambda x^2}{2l} = \frac{20 \times 0.6^2}{2 \times 1.2} = 3 \text{ J.}$$

Since the table is *smooth*, all of the EPE will be converted to KE as the string returns to its natural length and becomes slack.

$\text{KE} = \frac{1}{2}mv^2 = \frac{1}{2} \times 0.3 \times v^2 = 0.15v^2.$

This gives: $0.15v^2 = 3 \Rightarrow v = \sqrt{20}$

P has speed 4.47 ms^{-1} when the string becomes slack.

Progress check

1 An elastic string of natural length 0.8 m is extended to a length of 1.4 m by a force of 60 N. Find the modulus of elasticity of the string.

2 A light spring with modulus of elasticity 80 N and natural length 0.9 m is compressed by a distance of 0.2 m. One end of the spring is fixed and a particle of mass 0.1 kg is attached to the other end. The system is released from rest on a smooth horizontal surface.
Find the maximum speed reached by the particle in the subsequent motion.

1 80 N
2 5.96 ms^{-1} to 3 s.f.

4.3 Simple harmonic motion

After studying this section you should be able to:

- *understand the definition of SHM and the terms period and amplitude*
- *establish when linear oscillatory motion is simple harmonic and apply standard results to solve a problem*
- *use SHM to model the oscillation of a simple pendulum*

LEARNING SUMMARY

Defining simple harmonic motion

AQA	M2
EDEXCEL	M3
OCR	M3
WJEC	M3
NICCEA	M3

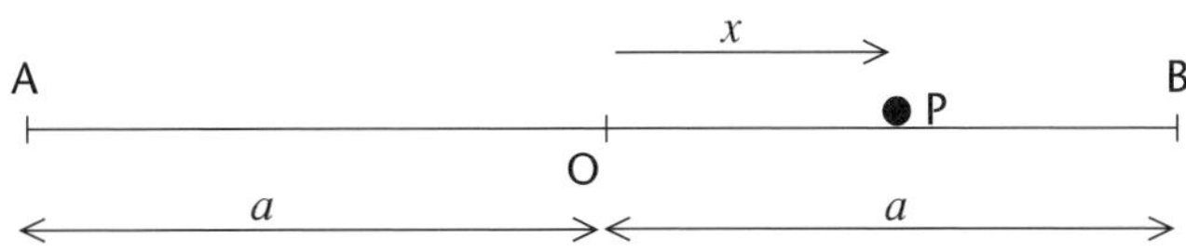

In the diagram, P is a particle that oscillates, in a straight line, between two extreme positions at A and B. O is at the centre of the oscillation and x is the displacement of P from O at some instant where $-a \leqslant x \leqslant a$ and a is called the **amplitude** of the motion. The time taken to complete one oscillation is called the **periodic time**.

So far, this just defines the movement of P as some kind of oscillation. To qualify as **simple harmonic motion**, **SHM**, the acceleration of P must always be proportional to x and be directed towards O.

This condition may be written concisely as a differential equation and several results may then be derived that apply to all cases of simple harmonic motion.

> P is said to move with SHM provided that $\ddot{x} = -\omega^2 x$ where ω is a constant. It then follows that:
>
> - the maximum acceleration has magnitude $\omega^2 a$ and occurs at the end-points where the speed is zero
> - $v = \omega\sqrt{a^2 - x^2}$ or $v^2 = \omega^2(a^2 - x^2)$
> - the maximum speed is ωa which occurs at O (when $x = 0$)
> - $x = a\sin(\omega t + \alpha)$
> - $T = \dfrac{2\pi}{\omega}$ where T is the period of the motion.
>
> KEY POINT

Example

A particle P is moving with linear SHM of amplitude 1.5 m. The speed of P is 5 ms^{-1} when its displacement from the centre of motion is 1 m.

Find:

(a) the maximum speed (b) the maximum acceleration (c) the periodic time.

(a) Using: $v^2 = \omega^2(a^2 - x^2)$

gives: $5^2 = \omega^2(1.5^2 - 1^2) \Rightarrow \omega = \sqrt{20} = 2\sqrt{5}$

so the maximum speed is $\omega a = 2\sqrt{5} \times 1.5 = 3\sqrt{5}$ ms^{-1}.

(b) The maximum acceleration is $\omega^2 a = 20 \times 1.5 = 30$ ms^{-2}.

(c) The periodic time is $T = \dfrac{2\pi}{\omega} = \dfrac{2\pi}{2\sqrt{5}} = \dfrac{\sqrt{5}\pi}{5}$ seconds.

Establishing linear SHM

AQA	M2
EDEXCEL	M3
OCR	M3
WJEC	M3
NICCEA	M3

KEY POINT

If you can establish that the oscillation of a particle is SHM then you can immediately apply the general results to find out more about how the particle will behave.
The key to establishing SHM is to show that the motion satisfies the equation:

$$\ddot{x} = -\omega^2 x.$$

Example

In the diagram, R moves anti-clockwise around the circle so that OR has constant angular speed ω radians/second. At time t, OR makes an angle ωt with the positive x-axis. P moves along the x-axis so that RP remains vertical.

Prove that P exhibits SHM.

From the diagram OP is given by $x = a \cos \omega t$

The chain rule is required here.

differentiating gives: $\dot{x} = -\omega a \sin \omega t$

differentiating again gives: $\ddot{x} = -\omega^2 a \cos \omega t$

so: $\ddot{x} = -\omega^2 x$

This proves that P oscillates with SHM.

Notice that this example provides a physical illustration of what the various constants in the equations may represent.

The simple pendulum

AQA	M2
EDEXCEL	M3
OCR	M3
WJEC	M3
NICCEA	M3

In the simple pendulum, the angle that the string makes with the vertical changes in a way that exhibits **angular SHM**.

Using $F = ma$ along the tangent to the path of the particle gives:

$$mg \sin \theta = -ml\ddot{\theta}$$

so: $\ddot{\theta} = -\frac{g}{l} \sin \theta$

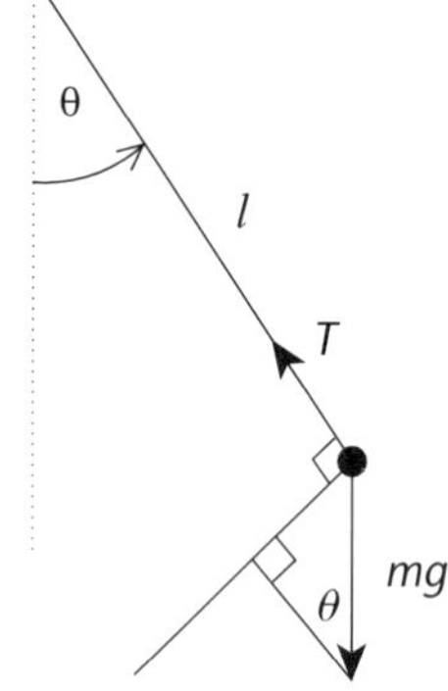

For small values of θ in radians, $\sin \theta \approx \theta$. This gives $\ddot{\theta} = -\frac{g}{l}\theta$ for small oscillations which represents SHM with $\omega = \sqrt{\frac{g}{l}}$.

T represents the periodic time here, whereas on the diagram it is used to represent the tension in the string.

Using $T = \frac{2\pi}{\omega}$ gives $T = 2\pi\sqrt{\frac{l}{g}}$.

Progress check

A particle, which describes linear SHM with O at the centre, has speed 5 ms^{-1} at a distance of 1 m from O and a maximum speed of 13 ms^{-1}. Find:

(a) the periodic time
(b) the amplitude of the motion
(c) the speed of the particle when it is 0.5 m from O.

(a) $\frac{\pi}{6}$ seconds (b) $\frac{13}{12}$ m (c) 11.5 ms^{-1}.

4.4 Further circular motion

After studying this section you should be able to:

- *solve problems involving uniform motion in a horizontal circle on banked tracks*
- *solve problems involving motion in a vertical circle*

LEARNING SUMMARY

Banked tracks

AQA	M3
EDEXCEL	M3
OCR	M2
WJEC	M3
NICCEA	M2

When a car travels in a horizontal circle, on a flat surface, the centripetal force required to keep it moving in a circle is supplied by friction. The speed of the car must be restricted because the frictional force has a limiting value. Once this limiting value is reached, any further increase in speed will cause the car to skid.

See p. 93 for a reminder about centripetal force.

One way to increase the maximum speed of a car in a horizontal circle is to bank the track so that the reaction force of the road surface on the car has a horizontal component. This can result in a greater value of the maximum centripetal force which allows greater speeds to be reached without skidding.

Example

A car travels around a track in a horizontal circle of radius 150 m.

(a) What angle does the track need to be banked at so that the car has no tendency to side-slip at a speed of 25 ms^{-1}?

(b) Given that $\mu = 0.7$ find the greatest speed that the car can travel around the track without slipping.

(a) There is no frictional force, in this case, since there is no tendency to slip

The forces balance vertically.

Vertically: $R \cos\theta = mg$ (1)

Using $F = ma$ where the acceleration is towards the centre of the circle.

Horizontally: $R \sin\theta = \dfrac{m \times 25^2}{150}$ (2)

(2) ÷ (1) gives: $\tan\theta = \dfrac{25^2}{150 \times 9.8}$

$\Rightarrow \theta = 23.0^0$

The track needs to be banked at 23.0^0.

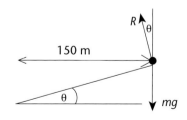

(b) Vertically: $R\cos 23^0 - 0.7R\sin 23^0 = mg$ (3)

Horizontally: $R\sin 23^0 + 0.7R\cos 23^0 = \dfrac{mv^2}{150}$ (4)

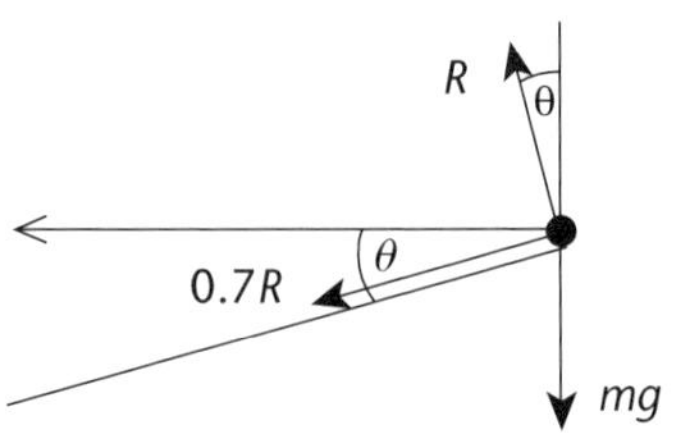

(4) ÷ (3) gives: $\dfrac{v^2}{150g} = \dfrac{\sin 23° + 0.7 \times \cos 23°}{\cos 23° - 0.7 \times \sin 23°}$

$= 1.5997 \ldots$

so: $v^2 = 1.5997 \ldots \times 150 \times 9.8$

$= 2351.76 \ldots \Rightarrow v = 48.49 \ldots$

The maximum speed is 48.5 ms^{-1} to 3 s.f.

Motion in a vertical circle

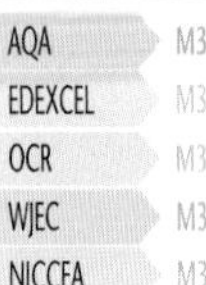

Solving problems regarding the motion of a particle in a vertical circle generally involves:

- calculation of the speed of the particle by considering energy
- consideration of the radial forces necessary to maintain circular motion.

KEY POINT

For a particle attached to a string it is assumed that the only forces acting on the particle are its weight and the tension in the string. Since the tension always acts in a direction perpendicular to the motion *it does no work* and so the total mechanical energy of the particle is conserved.

Example

A particle P of mass 0.3 kg is attached to one end of a light inelastic string of length 0.8 m. The other end of the string is attached to a fixed point O. The particle is given an initial speed of 5 ms^{-1} horizontally from a point 0.8 m vertically below O. Find:

(a) the speed of P when OP is horizontal
(b) the tension in the string at this point
(c) the angle that OP makes with the upward vertical when the string first becomes slack.

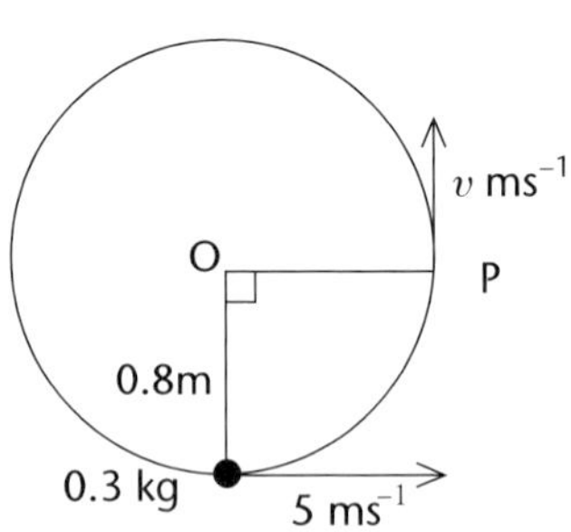

(a) Since energy is conserved:

$KE = \frac{1}{2}mv^2$
$GPE = mgh$

$\frac{1}{2} \times 0.3 \times v^2 = \frac{1}{2} \times 0.3 \times 5^2 - 0.3 \times 9.8 \times 0.8$

giving: $v^2 = 5^2 - 2 \times 9.8 \times 0.8 = 9.32$

$\Rightarrow v = 3.052 \ldots$

so the speed of P when OP is horizontal is 3.05 ms^{-1} to 3 s.f.

(b) Horizontally: $T = \dfrac{0.3 \times 9.32}{0.8} = 3.495$

The tension in the string is 3.50 N to 3 s.f.

(c) Using the conservation of energy principle:

$\frac{1}{2} \times 0.3 \times v^2 = \frac{1}{2} \times 0.3 \times 5^2 - 0.3 \times 9.8(0.8 + 0.8 \cos \theta)$

So $v^2 = 5^2 - 19.6 \times 0.8(1 + \cos \theta)$ (1)

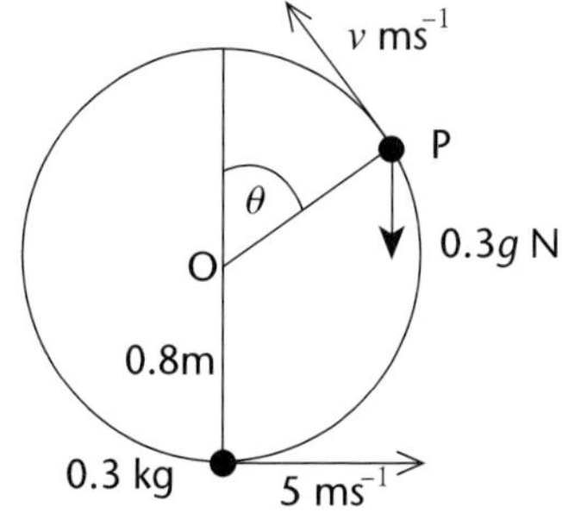

When the string first becomes slack, the component of weight along PO is equal to the force needed to maintain circular motion. This gives:

$$0.3 \times 9.8 \cos \theta = \frac{0.3v^2}{0.8} \Rightarrow v^2 = 0.8 \times 9.8 \cos \theta \qquad (2)$$

From (1) and (2) $\cos \theta = \frac{9.32}{23.52} \Rightarrow \theta = 66.7^0$

The string first becomes slack when OP makes an angle of 66.7^0 with the upward vertical.

Progress check

1 A car travels around a track in a horizontal circle of radius 200 m. What angle does the track need to be banked at so that the car has no tendency to side-slip at a speed of 30 ms^{-1}?

2 A particle P of mass 0.5 kg is attached to one end of a light inextensible string of length 0.6 m. The other end of the string is attached to a fixed point O. P is released from rest with OP horizontal and the string taut. Find the maximum tension in the string.

1 24.7^0
2 14.7 N

4.5 Centre of mass of uniform solids

After studying this section you should be able to:

- *use integration to find the centre of mass of a uniform lamina*
- *use integration to find the centre of mass of a solid of revolution*

LEARNING SUMMARY

Centre of mass of a uniform lamina

AQA M2
EDEXCEL M2
NICCEA M3

The method is to divide the lamina into thin strips of width δx to find an approximate expression for the position of the centre of mass. The exact position corresponds to the limit as δx approaches zero, and this defines an integral.

Example

The diagram shows a uniform lamina.
Find the coordinates of its centre of mass.

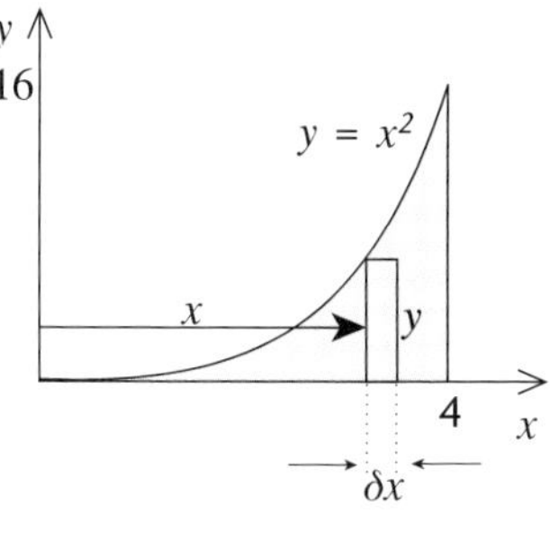

The mass of each strip is represented by its area $y\delta x$ and the total mass is represented by the area of the lamina.

See page 86 in the Mechanics 2 chapter for a reminder about the use of this formula.

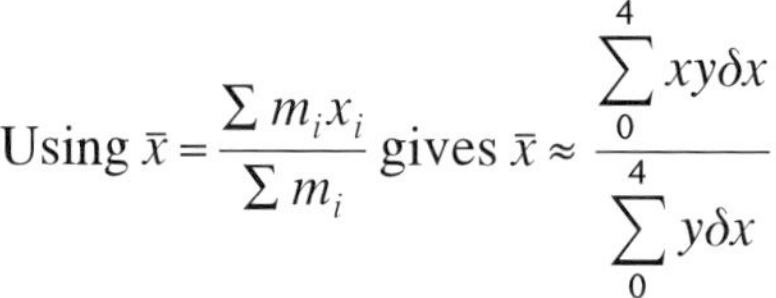

$$\text{Using } \bar{x} = \frac{\sum m_i x_i}{\sum m_i} \text{ gives } \bar{x} \approx \frac{\sum_0^4 xy\delta x}{\sum_0^4 y\delta x}$$

Substituting $y = x^2$ and converting to an integral

$$\text{gives the exact value as } \bar{x} = \frac{\int_0^4 x^3 \, dx}{\int_0^4 x^2 \, dx} = \frac{\left[\frac{x^4}{4}\right]_0^4}{\left[\frac{x^3}{3}\right]_0^4} = 3$$

The same process may be used to find the y-coordinate of the centre of mass. In this case, the distance x is replaced by $\frac{y}{2}$ which is the height of the centre of mass of each strip.

This gives:
$$\bar{y} \approx \frac{\sum_0^4 \frac{y}{2} \times y\delta x}{\sum_0^4 y\delta x}$$

which converts to:
$$\bar{y} = \frac{\frac{1}{2}\int_0^4 x^4 \, dx}{\frac{4^3}{3}}$$

$$= \frac{\frac{1}{2}\left[\frac{x^5}{5}\right]_0^4}{\frac{4^3}{3}}$$

$$= \frac{1}{2} \times \frac{4^2}{5} \times 3 = 4.8$$

The coordinates of the centre of mass are (3, 4.8).

The centre of mass of a uniform solid of revolution

AQA M2
EDEXCEL M3
NICCEA M3

The method used for a uniform solid of revolution is an extension of the one used for a lamina. In this case, the solid is divided into slices of thickness δx.
The mass of each slice is represented by its volume and the total mass is represented by the volume of the solid.

Example

Find the coordinates of the centre of mass of the uniform hemisphere shown in the diagram.

You can use the standard result for the volume of a hemisphere.

Using $\bar{x} = \dfrac{\Sigma m_i x_i}{\Sigma m_i}$ gives $\bar{x} \approx \dfrac{\sum_0^r \pi y^2 x \delta x}{\frac{2}{3}\pi r^3}$

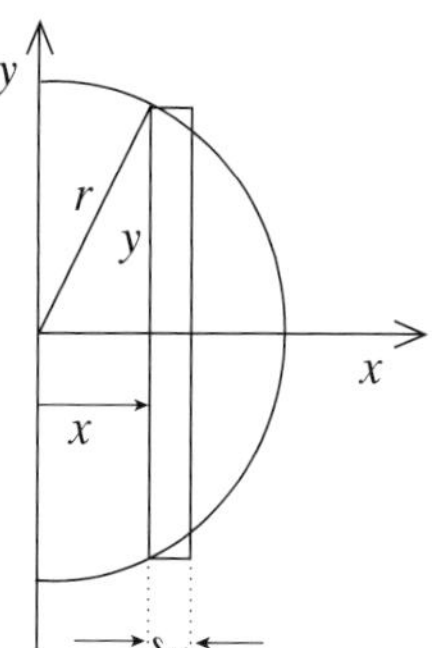

Substituting $y^2 = r^2 - x^2$ and converting to an integral gives the exact value as

$$\bar{x} = \frac{\int_0^r \pi(r^2 - x^2)x\,dx}{\frac{2}{3}\pi r^3}$$

$$= \frac{\int_0^r r^2x - x^3\,dx}{\frac{2}{3}r^3}$$

giving: $\bar{x} = \dfrac{\left[\dfrac{r^2x^2}{2} - \dfrac{x^4}{4}\right]_0^r}{\frac{2}{3}r^3} = \dfrac{r^4}{4} \times \dfrac{3}{2r^3}$

so: $\bar{x} = \dfrac{3r}{8}$. By symmetry, $\bar{y} = 0$.

The coordinates of the centre of mass are $\left(\dfrac{3r}{8}, 0\right)$

Progress check

1 Use integration to find the coordinates of the centre of mass of the uniform triangular lamina shown in the diagram.

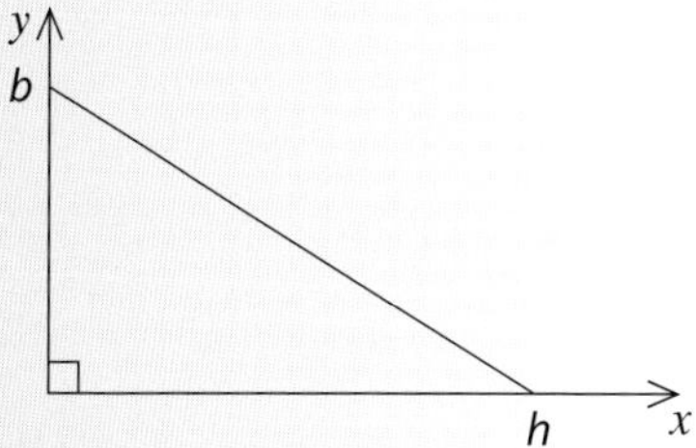

2 A solid of revolution is formed by rotating the graph of $y = \sqrt{x}$, between $x = 0$ and $x = 4$, through a full turn about the x-axis.
Use integration to find the coordinates of the centre of mass of the solid.

2 $\left(\frac{8}{3}, 0\right)$
1 $\left(\frac{h}{3}, \frac{b}{3}\right)$

4.6 Collisions in two dimensions

After studying this section you should be able to:

- *solve problems involving the oblique impact of two smooth spheres of equal size*

LEARNING SUMMARY

Oblique impact

OCR M3

The diagram shows two smooth spheres A and B of equal size, at the moment of impact. Regardless of which directions A and B were moving in before the impact, the contact forces lie along the line $C_A C_B$ through their centres. This means that:

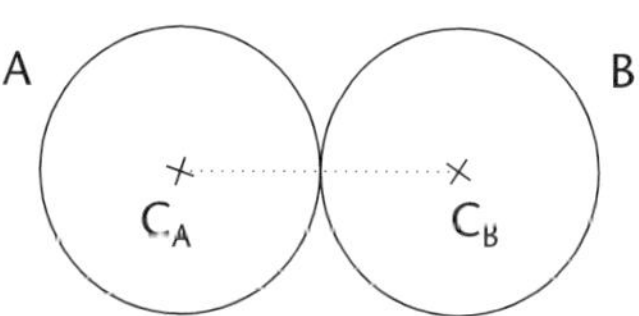

- there is no change in the component of the speed of either sphere in the direction perpendicular to $C_A C_B$
- along the line $C_A C_B$ the equations normally used for direct impact still apply.

Example

Sphere A is in collision with sphere B which is at rest. The spheres are of equal size but the mass of A is twice the mass of B. Before the impact, A is moving at an angle of 30^0 to their line of centres at 4 ms^{-1}. After the impact, A is moving at an angle of 60^0 to their line of centres. Find the speed of each sphere after the collision.

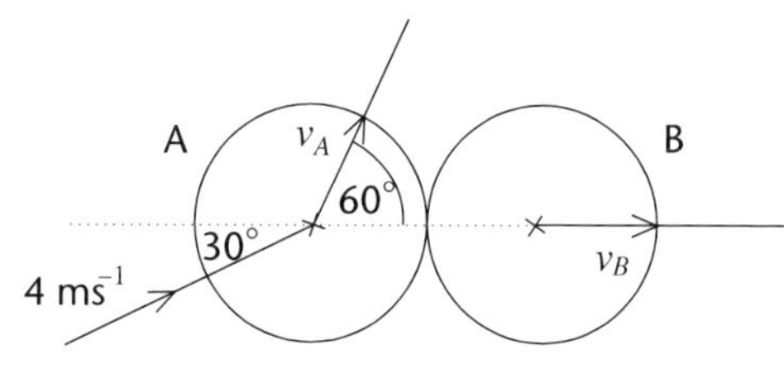

The component of A's velocity in this direction is unchanged by the impact.

Perpendicular to the line of centres: $v_A \sin 60^0 = 4 \sin 30^0$

$$v_A \times \frac{\sqrt{3}}{2} = 2 \Rightarrow v_A = \frac{4}{\sqrt{3}} = \frac{4\sqrt{3}}{3}$$

By the principle of conservation of momentum.

along the line of centres: $2m \times 4 \cos 30^0 = 2m \times \frac{4\sqrt{3}}{3} \cos 60^0 + mv_B$

this gives: $4\sqrt{3} = \frac{4\sqrt{3}}{3} + v_B$

$$\Rightarrow v_B = \frac{8\sqrt{3}}{3}$$

In this example, the direction of v_A was given in the question so the value of e wasn't needed.

After the impact, the speeds of A and B are $\frac{4\sqrt{3}}{3}$ ms^{-1} and $\frac{8\sqrt{3}}{3}$ ms^{-1} respectively.

Progress check

A sphere of mass 6 kg moving with speed 5 ms^{-1} strikes an equal-sized sphere of mass 4 kg which is at rest. The direction of motion is at 60^0 to the line of centres. After the collision, the 6 kg sphere moves at right angles to the line of centres.
Find the speed of each sphere after the collision.

1 6 kg sphere: $\frac{5\sqrt{3}}{2}$ ms^{-1}, 4 kg sphere: 3.75 ms^{-1}.

Sample questions and model answers

1

A light elastic string of natural length l and modulus of elasticity $3mg$ is fixed at one end and hangs vertically with a particle of mass m at the other end.

(a) Show that the extension of the string in the equilibrium position is $\frac{l}{3}$.

(b) The particle is then pulled down a further distance $\frac{l}{6}$ and released. Show that the particle oscillates with simple harmonic motion.

(c) Find the periodic time of the motion.

(d) Show that the maximum speed of the particle is $\frac{1}{2}\sqrt{\frac{gl}{3}}$.

It's a good idea to quote the formula that you use.

(a) Using $T = \frac{\lambda}{l}x$ gives $mg = \frac{3mg}{l}x \Rightarrow x = \frac{l}{3}$.

The extension in the string is $\frac{l}{3}$ as required.

Take a little time to set out a diagram labelled with all of the information.

(b) When the displacement of the particle below the equilibrium position is x, the tension in the string is given by:

Using $T = \frac{\lambda}{l}x$:

You can work from your diagram to fill in the details in the formula.

$$T = \frac{3mg}{l}\left(\frac{l}{3} + x\right) = mg + \frac{3mg}{l}x$$

The resultant upward force is $\frac{3mg}{l}x$.

Using $F = ma$, taking upwards as positive:

You need to establish an equation of the form $\ddot{x} = -\omega^2 x$

$$\frac{3mg}{l}x = -m\ddot{x} \Rightarrow \ddot{x} = -\frac{3g}{l}x$$

so the particle moves with SHM and $\omega = \sqrt{\frac{3g}{l}}$.

(c) The periodic time is given by $T = \frac{2\pi}{\omega} = 2\pi\sqrt{\frac{l}{3g}}$.

You may need to do some algebraic manipulation to establish the result in the required form.

(d) The maximum speed is given by $\omega a = \sqrt{\frac{3g}{l}} \times \frac{l}{6}$

$$= \frac{1}{2}\sqrt{\frac{3g}{l}} \times \frac{l}{3}$$

$$= \frac{1}{2}\sqrt{\frac{3g}{l} \times \frac{l^2}{9}}$$

$$= \frac{1}{2}\sqrt{\frac{gl}{3}}$$ as required.

Sample questions and model answers (continued)

2

A uniform cone has base radius r and height h.

(a) Show by integration that the centre of mass of the cone is at distance $\frac{h}{4}$ from its base.

(b) The cone is attached to one end of a cylinder of the same material. The cylinder has radius r and height h. Find an expression for the distance of the centre of mass of the composite shape from the base of the cylinder.

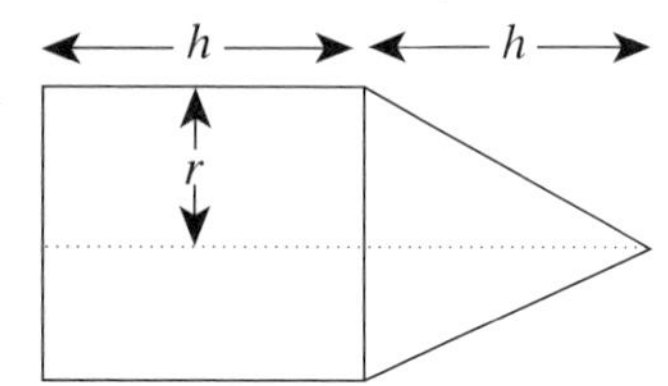

(a) The cone is shown with its axis of symmetry lying along the x-axis and its base in contact with the y-axis.

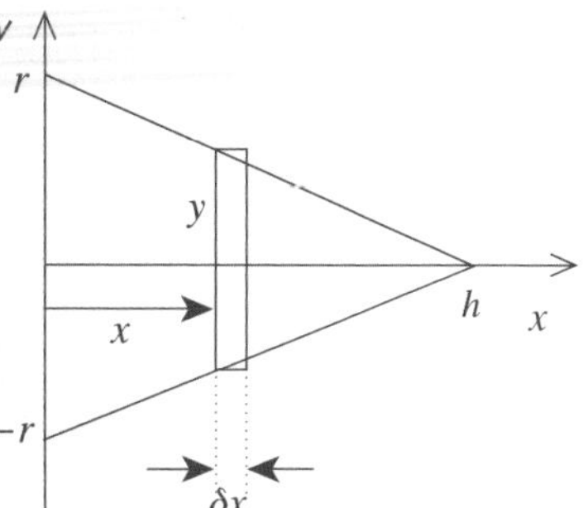

From the diagram $y = r - \frac{r}{h}x = r\left(1 - \frac{x}{h}\right)$.

The volume of a typical slice is $\pi r^2\left(1 - \frac{x}{h}\right)^2 \delta x$.

The cone is uniform so the volume of each slice may be used to represent its mass in the formula $\bar{x} = \frac{\sum m_i x_i}{\sum m_i}$. The total mass is represented by the volume of the cone.

This gives: $$\bar{x} \approx \frac{\sum_0^h \pi x r^2\left(1 - \frac{x}{h}\right)^2}{\frac{1}{3}\pi r^2 h} = \frac{3}{h}\sum_0^h x\left(1 - \frac{x}{h}\right)^2 \delta x$$

The exact value is: $$\bar{x} = \frac{3}{h}\int_0^h x\left(1 - \frac{x}{h}\right)^2 dx$$

$$= \frac{3}{h}\int_0^h x - \frac{2x^2}{h} + \frac{x^3}{h^2}\, dx = \frac{3}{h}\left[\frac{x^2}{2} - \frac{2x^3}{3h} + \frac{x^4}{4h^2}\right]_0^h$$

$$= \frac{3}{h}\left(\frac{h^2}{2} - \frac{2h^2}{3} + \frac{h^2}{4}\right) = 3h\left(\frac{1}{2} - \frac{2}{3} + \frac{1}{4}\right) = 3h \times \frac{1}{12}$$

giving $\bar{x} = \frac{h}{4}$ as required.

(b) The volume of the cone is $\frac{1}{3}\pi r^2 h$.
The volume of the cylinder is $\pi r^2 h$.
The ratio of their volumes is 1 : 3.

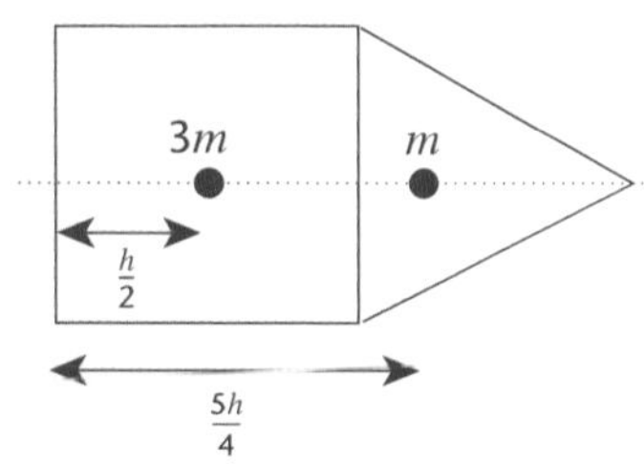

For the composite shape, this gives:

$$\bar{x} = \frac{3m \times \frac{h}{2} + m \times \frac{5h}{4}}{4m} = \frac{\frac{11mh}{4}}{4m} = \frac{11h}{16}$$

The centre of mass of the composite shape is $\frac{11h}{16}$ from its base.

Practice examination questions

1 A particle P moves along a straight line passing through the point O. The displacement x metres, of the particle from O at time t seconds is given by:

$x = 4t^3 - 3t^2 + 7$ for $0 \leqslant t \leqslant 2$.

Find:

(a) the velocity of the particle at time t

(b) the distance that the particle travels before it changes direction

(c) the time taken for the particle to return to its starting point.

2 A particle moving in a straight line is acted on by a single retarding force with magnitude proportional to the square of its speed. The initial speed of the particle is 20 ms^{-1} and its initial retardation is 3 ms^{-2}.

(a) Show that the speed v ms^{-1} of the particle after t seconds is given by:

$$v = \frac{400}{3t + 20}$$

(b) Given that the particle has mass 2 kg, find the magnitude of the work done by the retarding force between $t = 0$ and $t = 4$.

3 A light elastic string of natural length 2 m and modulus of elasticity $5mg$ has one end fixed at a point O. The other end is attached to an object of mass m which is released from rest at O.

Find:

(a) the distance that the object falls below O before being brought to instantaneous rest by the tension in the string

(b) the acceleration of the object at its lowest point.

4 A particle of mass 0.5 kg is attached to two springs of natural length 0.8 m and modulus of elasticity 10 N. The springs lie in a straight line on a smooth horizontal table with their ends fixed at points A and B 1.6 m apart.

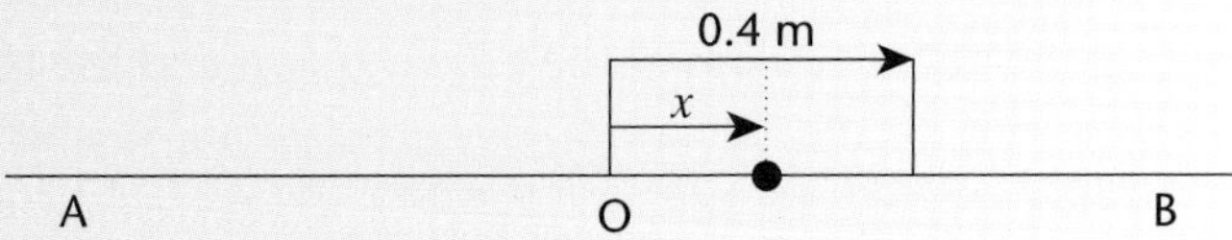

The particle is then displaced 0.4 m towards B and released. Given that the displacement of the particle from the equilibrium position after t seconds is x metres:

(a) show that $\ddot{x} = -50x$

(b) find the time required to complete one oscillation to 1 d.p.

(c) find the average speed of the particle during the first 0.3 seconds to 1 d.p.

Practice examination questions (continued)

5 A car is driven round a track in a horizontal circle of radius 80 m at a speed of 20 ms^{-1}. The track is banked at angle α to the horizontal and there is no tendency for the car to side-slip.

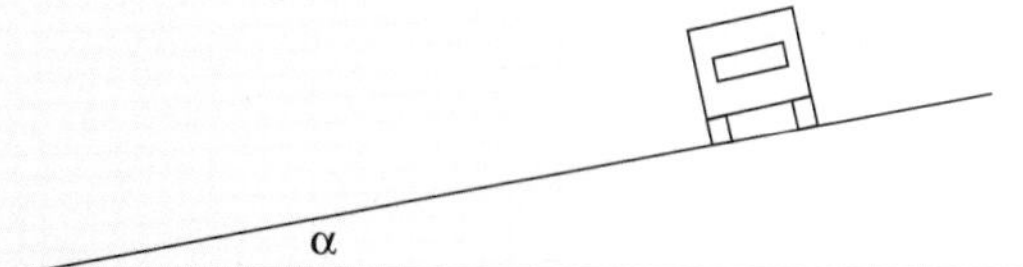

By representing the car as a particle:

(a) Find α to 1 d.p.

(b) The maximum speed that the car can travel on the banked track without slipping sideways is 28 ms^{-1}. Find the coefficient of friction between the tyres and the surface of the track.

6 The diagram shows a particle of mass 0.2 kg attached to one end of a light inelastic string of length 80 cm.

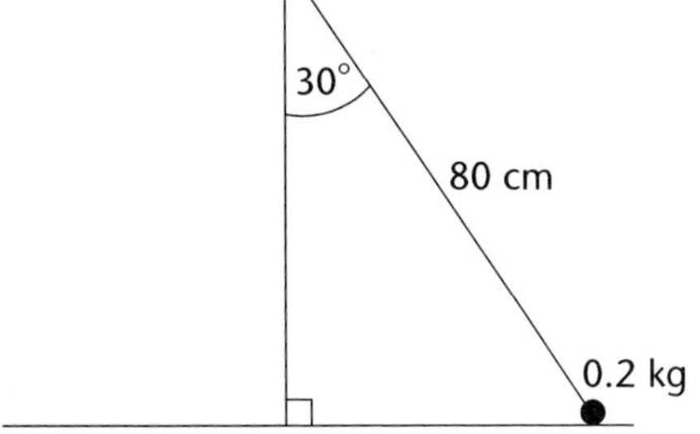

The particle moves in a horizontal circle on a smooth surface with the string inclined at 30° to the vertical.

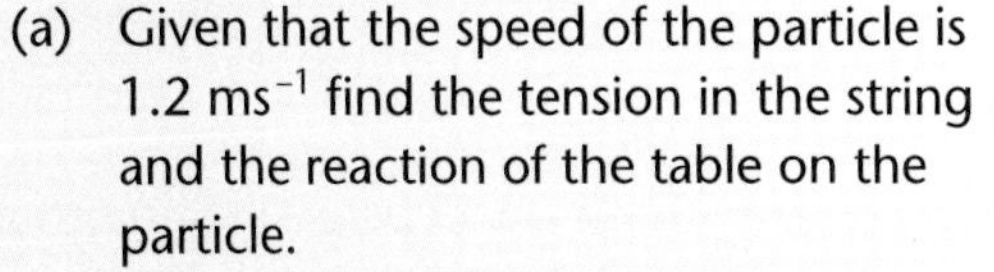

(a) Given that the speed of the particle is 1.2 ms^{-1} find the tension in the string and the reaction of the table on the particle.

(b) Find the minimum angular speed of the particle so that the reaction of the table on the particle is zero.

7 A small ball B is released from rest at the top of a horizontal circular tube. The ball travels down the outside of the tube's surface following a circular path, with centre C, until it falls away.

Given that the outside radius of the tube is 0.6 m and the mass of the ball is 0.1 kg find:

(a) an expression for the speed of B when CB makes an angle θ with the upward vertical

(b) the reaction of the tube on the ball in terms of θ and g

(c) the value of θ when the ball loses contact with the surface of the tube. Take $g = 9.8$ ms^{-2}.

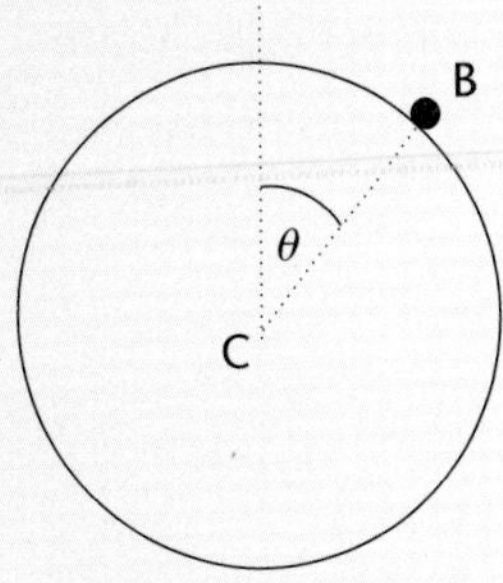

Practice examination questions (continued)

8 (a) The diagram shows a uniform lamina in the shape of a quadrant.

Use integration to find the coordinates of its centre of mass.

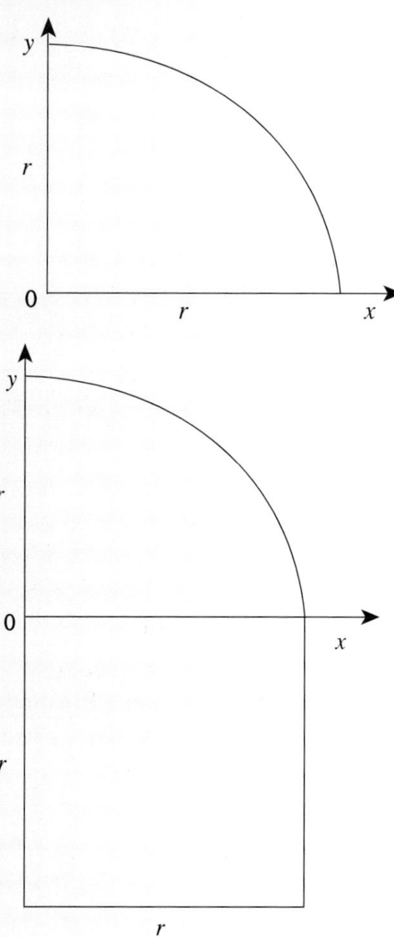

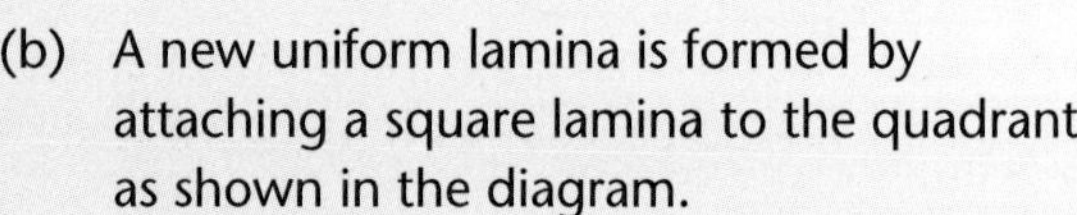

(b) A new uniform lamina is formed by attaching a square lamina to the quadrant as shown in the diagram.

Show that the x – coordinate of the centre of mass of this lamina is given by:

$$\frac{10r}{3\pi + 12}.$$

Chapter 5

Decision and discrete mathematics 2

The following topics are covered in this chapter:

- Game theory
- Recurrence relations
- Coding

5.1 Game theory

After studying this section you should be able to:

- understand the idea of a zero-sum game and its representation using a pay-off matrix
- identify play-safe strategies and stable solutions
- find optimal mixed strategies for a game with no stable solution
- reduce a pay-off matrix using a dominance argument

LEARNING SUMMARY

A two person zero-sum game

EDEXCEL D2
OCR D2

The starting point in the mathematical theory of games is that the outcome of a game is determined by the **strategies** of the players.

A two person **zero-sum game** is a game in which the winnings of one player equal the losses of the other for every combination of strategies. Taking winnings to be positive and losses to be negative gives a zero sum in each case.

Viewing a game from one player's point of view, we could represent the outcomes (called pay-offs) for each combination of strategies in a matrix. This is called the **pay-off matrix** for that player.

Example

A and B are two players in a zero-sum game. A uses one of two strategies, W or X, and B uses one of the strategies Y or Z. The table shows the pay-off matrix for A.

The situation could equally be represented by the pay-off matrix for B. This would show corresponding values with opposite signs since this is a zero-sum game.

Pay-off matrix for A		B: Y	B: Z
A	W	2	–2
A	X	5	–4

The pay-off matrix shows that if B adopts strategy Y then the pay-off for A will be 2 by using strategy W and 5 using strategy X. On the other hand, if B adopts strategy Z then the pay-off for A will be –2 using strategy W and –4 using strategy X.

The idea is that neither player knows in advance which strategy the other will use.

The play-safe strategy

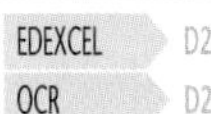
EDEXCEL D2
OCR D2

The **play-safe** strategy for a player is the strategy for which the minimum pay-off is as high as possible. In the example above, the minimum pay-off for A using strategy W is −2, whereas the minimum pay-off using strategy X is −4. This means that strategy W is the play-safe strategy for player A.

Notice that finding the play-safe strategy for player A involves comparing the minimum values in the *rows* of the pay-off matrix for A. Finding the play-safe strategy for B will involve comparing the values in the *columns*, remembering that B's pay-offs are the negatives of the ones in the pay-off matrix for A.

The minimum value for B using strategy Y is −5 and the minimum value using strategy Z is 2. This means that the play-safe strategy for B is strategy Z.

The situation is shown in the pay-off matrices for A and B.

Pay-off matrix for A		B: Y	B: Z
A	W	2	−2
A	X	5	−4

Pay-off matrix for B		B: Y	B: Z
A	W	−2	2
A	X	−5	4

In this case, the maximum of the minimum pay-offs, for each player, is in the corresponding position in the two matrices. This represents the **stable solution** to the problem referred to as the **saddle point** (or **minimax point**).

The solution is stable in the sense that neither player can improve their pay-off by taking a different strategy, given that the other player doesn't change. In other words, while B uses strategy Z, the best strategy for A is W and while A uses strategy W, the best strategy for B is Z.

In the example, the sum of these values is −2 + 2 = 0 and the solution is stable.

KEY POINT

If the sum of the two values used to determine the play-safe strategies is **not zero** then the values cannot correspond to the same cell position in the play-off matrices. This means that there is no saddle point and the game has **no stable solution**.

Example

The pay-off matrices for two players in a zero-sum game are given below. Show that there is no stable solution for the game.

Pay-off matrix for P		Q: Y	Q: Z
P	W	5	−3
P	X	−4	6

Pay-off matrix for Q		Q: Y	Q: Z
P	W	−5	3
P	X	4	−6

The play-safe strategies for P and Q are shown shaded.
The values used to determine the play-safe strategies are −3 and −5.
There is no stable solution since $(-3) + (-5) \neq 0$.

Optimal strategies for games that are not stable

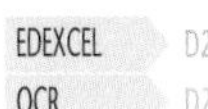
EDEXCEL D2
OCR D2

The repeated use of the same strategy over a series of games is called a **pure strategy**. This provides the best results for both players in a game which has a stable solution. In the case where no stable solution exists, a **mixed strategy** is used in which each of the strategies is employed with a given probability to find the optimal solution.

Returning to the previous example:

Pay-off matrix for P		Q: Y	Q: Z
P	W	5	–3
P	X	–4	6

Suppose that P chooses:
strategy W with probability p
and strategy X with probability $(1-p)$.

Then if:
Q chooses strategy Y, the expected *gain* for P is given by $5p - 4(1 - p) = 9p - 4$
Q chooses strategy Z, the expected *gain* for P is $-3p + 6(1 - p) = -9p + 6$

The optimal value occurs when these expressions are equal.

$$9p - 4 = -9p + 6 \Rightarrow 18p = 10$$
$$\Rightarrow p = \tfrac{5}{9}$$

The value of the game is given by $9p - 4 = 1$.

You could equally use $-9p + 6$ to get the value of the game.

Suppose that Q chooses:
strategy Y with probability q
and strategy Z with probability $(1 - q)$.

Then if:
P chooses strategy W, the expected *loss* for Q is given by $5q - 3(1 - q) = 8q - 3$
P chooses strategy X, the expected *loss* for Q is $-4q + 6(1 - q) = -10q + 6$

$$8q - 3 = -10q + 6 \Rightarrow 18q = 9$$
$$\Rightarrow q = \tfrac{1}{2}$$

(As a check, the value of the game is $8q - 3 = 1$ as before).

This shows that the optimal strategy for both players is to use mixed strategies such that:

P chooses strategy W with probability $\frac{5}{9}$ and strategy X with probability $\frac{4}{9}$

Q chooses strategy Y and strategy Z with equal likelihood.

The expected long-term gain then for P, as an average per game, is 1 and this is also the expected long-term loss, as an average per game, for Q.

Graphical representation

Each graph corresponds to a strategy for Q (i.e. the opponent of P).

The diagram shows graphs of $9p - 4$ and $-9p + 6$ against values of p from 0 to 1.

The point of intersection corresponds to the probability that gives the optimal mixed strategy for P in the last example.

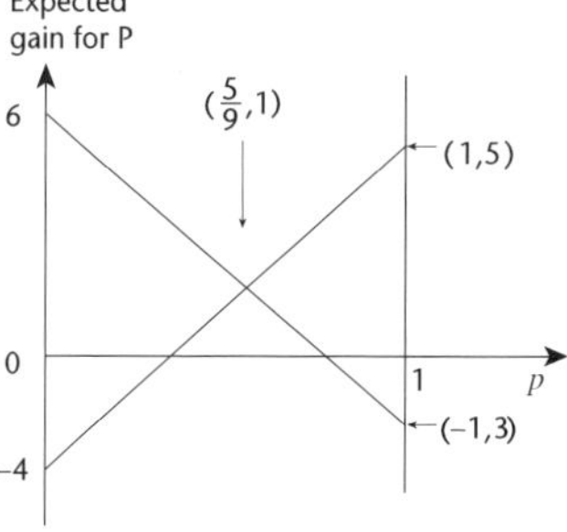

KEY POINT

When P's opponent has more strategies there will be more graphs with several points of intersection. You will need to identify the one that represents the optimal mixed strategy for P.

This will be the highest point on or below each of the graphs.

This diagram represents a situation where P's opponent has three strategies to choose from.

The point representing the optimal mixed strategy for P is circled.

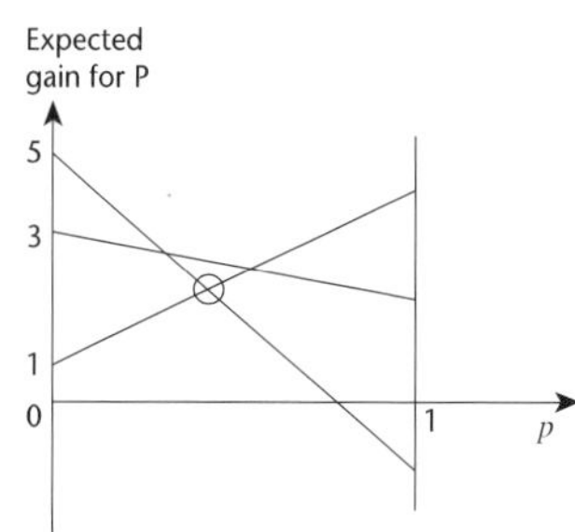

Notice how the problem of identifying the right vertex can be expressed as a linear programming problem in which the object is to maximise the expected gain for P subject to the constraints represented by the regions bounded by the straight line graphs.

See the Sample examination questions on p. 126.

This is, in fact, the approach used for higher dimensional problems. The conditions are formulated as a linear programming problem which may then be solved by the simplex algorithm.

See Practice examination questions on p. 128.

KEY POINT

If the pay-offs for one strategy are *always* better than the corresponding pay-offs for some other strategy then the weaker one can be ignored when determining the probabilities for a mixed strategy. In this way, the pay-off matrix is reduced by what is called a **dominance argument**.

Progress check

The table shows a pay-off matrix for player A in a zero-sum game.

(a) Find the play-safe strategy for each player.
(b) Show that this game has no stable solution.
(c) Find the best strategy for each player and the value of the game.

Pay-off matrix for A		Q	
		Y	Z
A	W	6	−4
	X	−3	5

(a) A(strategy X), Q(strategy Z)
(b) $(-3) + (-5) \neq 0$
(c) A Use W and X with probabilities $\frac{4}{9}$ and $\frac{5}{9}$
Q Use Y and Z with equal probability, value of game = 1

5.2 Recurrence relations

After studying this section you should be able to:

- *solve first order linear recurrence relations*
- *solve second-order homogeneous recurrence relations*
- *solve second-order non-homogeneous recurrence relations*

LEARNING SUMMARY

First-order linear recurrence relations

AQA D2

You need to be able to solve **first-order linear recurrence relations** of the form $x_n = ax_{n-1} + b$ where a and b are constants.

If, for example, the value of x_0 is known then the value of x_1 may be found using

$x_1 = ax_0 + b.$

The value of x_2 may then be found from the value of x_1 using

$x_2 = ax_1 + b.$

Continuing this process produces a sequence of **iterations** $x_1, x_2, x_3, \ldots$ starting from the value x_0 (called the **initial condition**).

One way to find the value of x_n for a particular value of n is to work out all of the iterations up to that value. Another way is to find a general expression for x_n in terms of n, called the **general solution**, and to substitute the known values.

In the case where $b = 0$, the relationship between successive iterations is given by $x_n = ax_{n-1}$. This gives:

$$x_1 = ax_0$$
$$x_2 = a^2x_0$$
$$x_3 = a^3x_0$$

If the sequence starts with x_1 then the process may be adapted to give the general solution as $x_n = a^{n-1}x_1$

and so the **general solution** is given by $x_n = a^n x_0$.

This may now be used to find a **particular solution** for some given initial condition.

In the case where $b \neq 0$, the iterations are:

$$x_1 = ax_0 + b$$
$$x_2 = a^2x_0 + ab + b$$
$$x_3 = a^3x_0 + a^2b + ab + b$$

Continuing in the same way gives:

The expression in the brackets is a G.P.

$$x_n = a^n x_0 + a^{n-1}b + a^{n-2}b + \ldots + a^3b + a^2b + ab + b$$
$$= a^n x_0 + b(1 + a + a^2 + a^3 + \ldots a^{n-1})$$

and so the **general solution** is given by $x_n = a^n x_0 + b\dfrac{a^n - 1}{a - 1}$ where $a \neq 1$.

Example

Find the general solution of the recurrence relation $x_n = 1.2x_{n-1} + 3$ and find the particular solution given that $x_0 = 10$.

The general solution is given by $x_n = 1.2^n x_0 + 3\dfrac{1.2^n - 1}{1.2 - 1} = 1.2^n x_0 + 15(1.2^n - 1)$.

The particular solution with $x_0 = 10$ is given by $x_n = 10 \times 1.2^n + 15(1.2^n - 1)$
$= 25 \times 1.2^n - 15.$

Second-order linear recurrence relations

AQA D2

You need to be able to deal with **second-order linear recurrence relations** of the form $x_n = ax_{n-1} + bx_{n-2} + p$ where $n \geqslant 2$ and a and b are constants. p is either zero or takes the form $cn + d$ or k^n. In the case where $p = 0$, the recurrence relation is said to be **homogeneous.**

In the exam, the auxiliary equation will always have real roots.

The first step in solving a homogeneous linear recurrence relation is to find and solve the **auxiliary equation**. The auxiliary equation for $x_n = ax_{n-1} + bx_{n-2}$ is the quadratic equation $m^2 - am - b = 0$.

If the auxiliary equation has two distinct roots m_1 and m_2 then the general solution of the recurrence relation is $x_n = Am_1^n + Bm_2^n$

For example, the auxiliary equation for $x_n = 8x_{n-1} - 15x_{n-2}$ is $m^2 - 8m + 15 = 0$. The roots are 3 and 5 so the general solution of the recurrence relation is $x_n = A3^n + B5^n$

If the auxiliary equation has equal roots (m) then the general solution of the recurrence relation becomes $x_n = Am^n + Bnm^n$

For example, the auxiliary equation for $x_n = 10x_{n-1} - 25x_{n-2}$ is $m^2 - 10m + 25 = 0$. The only root this time is 5 so the general solution of the recurrence relation is $x_n = A5^n + Bn5^n$

In the exam you will be given clear guidance about finding a particular solution.

To solve a **non-homogeneous linear recurrence relation** you need to solve the associated homogeneous part as above and then find a particular solution. The general solution takes the form:

x_n = (solution of associated homogeneous part) + (particular solution).

Example

Find the general solution of $x_n - x_{n-1} - 6x_{n-2} = 12n - 32$ given that a particular solution is of the form $an + b$.

The auxiliary equation is $m^2 - m - 6 = 0$. This has roots −2 and 3.
So, the general solution of $x_n - x_{n-1} - 6x_{n-2} = 0$ is $x_n = A(-2)^n + B3^n$.

Substituting $x_n = an + b$ in the original recurrence relation gives:

$an + b - (a(n-1) + b) - 6(a(n-2) + b) = 12n - 32.$

Verify these results for yourself.

Equating coefficients of n gives $a = -2$. Equating constant terms gives $b = 1$. A particular solution is $x_n = -2n + 1$.

The general solution of the recurrence relation is $x_n = A(-2)^n + B3^n - 2n + 1$.

Progress check

1 Find the general solution of $x_n = 1.5x_{n-1} + 3$ and the particular solution when $x_0 = 5$.

2 Find the general solution of $x_n = -3x_{n-1} + 10x_{n-2}$ and the particular solution when $x_0 = 4$ and $x_1 = 1$.

1 $x_n = 1.5^n x_0 + 6(1.5^n - 1)$, $x_n = 11 \times 1.5^n - 6$.
2 $x_n = A(-5)^n + B(2)^n$, $x_n = (-5)^n + 3(2)^n$

5.3 Coding

After studying this section you should be able to:

- *understand what a binary code is*
- *understand how to detect and correct errors in transmitting binary codes*
- *know the meaning of the Hamming distance and how it is used*
- *understand what linear codes are*
- *use parity check matrices to detect errors*

LEARNING SUMMARY

Binary codes

A **binary code** is a code represented by a two-state system. For example, a line of light bulbs can represent a binary code because each bulb can be in one of two states, on or off.

Morse code uses combinations of long and short beeps to represent the letters of the alphabet and so it is a binary code. The familiar bar-codes seen on items in a supermarket use combinations of thick and thin lines as a binary code to represent product information. The spaces in between the lines can also be thick or thin and form part of the code.

Any binary code can be represented numerically using combinations of 1s and 0s. Each 0 or 1 is a **binary digit** or **bit** and these are used to make a **codeword**. A particular **code** will be made up of a number of codewords of a given **length**. For example, the code {000, 001, 010, 100} is made up of four codewords of length three.

Detecting errors

Errors may occur during the transmission of a code, for example over the internet. One method of **error detection** is to use an extra bit at the end of each codeword as a **parity check**. The idea is to keep the total number of 1s in the codeword *even* so an extra 1 is used if the original codeword has an odd number of 1s, otherwise the extra bit is 0.

Example

The following codewords include a parity check bit in each case. Which ones indicate that there is an error?

10111 11010 00101 10011

The codewords where an error will be detected by this method will have an odd number of 1s.

These are: 11010 and 10011.

The method is most effective when there is the likelihood of single errors being contained in a codeword. If there are two errors, for example, then the parity check will fail to recognise any error.

Hamming distance

The hamming distance between two codewords is the number of places in which their bits differ. For example, the codewords 1010101 and 1110001 differ in two places so the hamming distance between them is 2.

The concept of distance is important in **error correction**. It is assumed that when a codeword is received that contains an error, the codeword *closest* to the one received, *from the list of codewords defining the code*, is the one intended.

It makes sense to keep the minimum hamming distance between any pair of different codewords in a code large enough so that, even when a received codeword has more than one error, the correct codeword is still closest to the one received.

For example, the minimum distance between different codewords in the code C = {00000, 00111, 11010, 11101} is 3. In this case, provided that a received codeword has no more than one error, it will be possible to identify the correct codeword. So, if the received codeword is 00011 for example, it may be taken that the intended codeword is 00111. The minimum distance has to be 5 or more in order to cope with the possibility of two errors.

In general, for a code to be able to deal correctly with up to e errors, the minimum hamming distance for the code must be at least $2e + 1$.

Linear codes

AQA D2

Using the rules $1 + 0 = 0 + 1 = 1$, $0 + 0 = 0$ and $1 + 1 = 0$ it is possible to add two codewords together by adding the corresponding bits.
For example, $1101 + 1011 = 0110$.

If the sum of any two codewords from a code always belongs to the code then the code is said to be **linear**.

> **KEY POINT**
> Every linear code must contain a codeword with a zero in each place since adding a codeword to itself will produce a codeword of equal length made entirely of zeros.

A **parity check matrix M** may be found for a linear code such that if c is any codeword belonging to the code then $\mathbf{M}c^{\mathrm{T}} = 0$.

If, for some received codeword r, $\mathbf{M}r^{\mathrm{T}} \neq 0$ then the resulting vector is called the **error syndrome** of r. This indicates an error in r corresponding to the position of the error syndrome in **M**.

For example, the linear code {0000, 1100, 0011, 1111} has parity matrix $\mathbf{M} = \begin{pmatrix} 1 & 1 & 0 & 0 \\ 1 & 1 & 1 & 1 \end{pmatrix}$. If the codeword $r = 1011$ is received then using **M** gives:

Note that 0 is used here to represent a column matrix in which each element is 0.

$\begin{pmatrix} 1 & 1 & 0 & 0 \\ 1 & 1 & 1 & 1 \end{pmatrix}\begin{pmatrix} 1 \\ 0 \\ 1 \\ 1 \end{pmatrix} = \begin{pmatrix} 1 \\ 1 \end{pmatrix}$. This shows that there is an error since it is non-zero.

c^{T} represents the transpose of the vector c.

Since the error syndrome matches the first two columns of **M**, it suggests that the error is in either the first or second bit of r. The correct codeword would then either be 0011 or 1111. In this case the result is still ambiguous but this is the best that can be expected because the minimum hamming distance for the code is so small.

The parity check matrix method will not be able to correct a received codeword if it contains more than one error.

Progress check

1 Find the minimum hamming distance required between two different codewords within a code in order that up to 3 errors may be corrected.

2 Show that the code {0000, 1010, 1110, 1000} is not linear.

1 7

2 For example, $1010 + 1110 = 0100$ and this does not belong to the code.

Sample questions and model answers

1

Ashley and Emma play a two-person zero-sum game in which they each consider three possible strategies. The table shows the pay-off matrix for Ashley.

		Emma		
		X	Y	Z
Ashley	A	5	−3	3
	B	4	1	−2
	C	−3	6	−4

(a) Find the play-safe strategy for each player and show that the game does not have a stable solution.

(b) Set up a linear programming problem to find Ashley's optimal mixed strategy. You are not required to solve it.

(a) The minimum row values are −3, −2 and −4 in Ashley's pay-off matrix. The maximum of these is −2 and so the pay-safe strategy for Ashley is strategy B.

The minimum column values in Emma's pay-off matrix are −5, −6 and −3. The maximum of these is −3 and so the play-safe strategy for Emma is strategy Z.

		Emma		
		X	Y	Z
Ashley	A	5	−3	3
	B	4	1	−2
	C	−3	6	−4

		Emma		
		X	Y	Z
Ashley	A	−5	3	−3
	B	−4	−1	2
	C	3	−6	4

The values use to determine the play-safe strategies do not have the corresponding positions in the pay-off matrices so there is no stable solution. Alternatively, $-3 + -2 \neq 0$ and so, again, there is no stable solution.

Sample questions and model answers (continued)

The values need to be made positive by adding a fixed value to each one.

(b) Adding 5 to each of the figures in Ashley's pay-off matrix gives:

10	2	8
9	6	3
2	11	1

Using a, b and c to represent the probability that Ashley uses strategy A, B or C respectively, the linear programming problem is formulated as:

Maximise P subject to $a+b+c \leqslant 1$ where $a \geqslant 0$, $b \geqslant 0$, $c \geqslant 0$ and

$$P - 10a - 9b - 2c \leqslant 0$$
$$P - 2a - 6b - 11c \leqslant 0$$
$$P - 8a - 3b - c \leqslant 0$$

and P represents the value of the game.

2

The equation is non-homogeneous.

Find the general solution of the recurrence relation $x_n - 7x_{n-1} + 12x_{n-2} = 5^n$ given that a particular solution is of the form $c5^n$.

Solve the associated homogeneous equation first using the auxiliary equation.

The associated homogeneous recurrence relation is $x_n - 7x_{n-1} + 12x_{n-2} = 0$.

This has auxiliary equation $m^2 - 7m + 12 = 0$

$$\Rightarrow \quad (m-3)(m-4) = 0$$
$$\Rightarrow \quad m = 3 \text{ or } m = 4.$$

So, the general solution of $x_n - 7x_{n-1} + 12x_{n-2} = 0$ is

$x_n = A3^n + B4^n$ for some constants A and B.

Substitute the form of the particular solution to produce an equation from which c can be found.

Substituting $x_n = c5^n$ in the original recurrence relation gives:

$$c5^n - 7c5^{n-1} + 12c5^{n-2} = 5^n$$
$$\Rightarrow \quad 5^2c - 7c \times 5 + 12c = 5^2$$
$$\Rightarrow \quad 25c - 35c + 12c = 25$$
$$\Rightarrow \quad 2c = 25$$
$$\Rightarrow \quad c = 12.5$$

So, a particular solution is $x_n = 12.5 \times 5^n$

The general solution is the sum of the general solution of the associated homogeneous equation and a particular solution.

The general solution is then $x_n = A3^n + B4^n + 12.5 \times 5^n$

Practice examination questions

1 A and B are the players in a two-person zero-sum game. Each may use one of two strategies when a game is played. The pay-off matrix for A is shown below.

		Player B	
Player A		Y	Z
	W	4	8
	X	6	3

(a) Find the best mixed strategy for each player.

(b) Find the value of the game.

2 Richard and Judy play a zero-sum game. The pay-off matrix for Richard is shown in the table.

		Judy		
		X	Y	Z
Richard	A	4	3	−4
	B	−3	6	8
	C	2	1	−6

(a) Find the play-safe strategy for each player.

(b) Show that there is no saddle point for the game.

(c) Explain why strategy C will not be part of Richard's mixed strategy.

(d) Find Richard's optimal mixed strategy.

3 Jack takes out a loan of £15 000 over 5 years. He makes a payment of £p every month which includes interest at the rate of 0.8% per month on the reducing balance.

(a) Write a first order linear recurrence relation to show how the amount owing at the end of a given month depends on the amount owing at the end of the previous month.

(b) Find the general solution of the recurrence relation.

(c) Find the value of p.

4 (a) Write down the auxiliary equation for the associated homogeneous recurrence relation of $x_n - 3x_{n-1} - 28x_{n-2} = 15n - 36$.

(b) Solve the auxiliary equation and write down the corresponding general solution for the homogeneous recurrence relation.

(c) Find the values of a and b such that $an + b$ is a particular solution of $x_n - 3x_{n-1} - 28x_{n-2} = 15n - 36$.

(d) Write down the general solution of $x_n - 3x_{n-1} - 28x_{n-2} = 15n - 36$.

Practice examination questions (continued)

5 The code C is defined as {0000, 1010, 0101, 1111}.

(a) Find the minimum hamming distance between codewords in C.

(b) Show that C is a linear code.

(c) Find a suitable parity matrix for C.

6 (a) Show that the matrix $\mathbf{M} = \begin{pmatrix} 0 & 1 & 0 & 1 & 0 \\ 1 & 0 & 1 & 0 & 0 \\ 1 & 0 & 0 & 0 & 1 \end{pmatrix}$ is a parity matrix for the code

{00000, 10101, 01010, 11111}.

(b) Use **M** to correct the received codeword 01011 given that it only contains one error.

Practice examination answers

Pure 3

1
$$\frac{1}{(1+x)^2} = (1+x)^{-2}$$
$$= 1 + (-2)x + \frac{(-2)(-3)}{2!}x^2 + \ldots$$
$$= 1 - 2x + 3x^2 + \ldots \quad |x| < 1$$
$$\sqrt{4+x} = \sqrt{4\left(1+\frac{x}{4}\right)}$$
$$= 4^{\frac{1}{2}}\left(1+\frac{x}{4}\right)^{\frac{1}{2}}$$
$$= 2\left(1+\frac{x}{4}\right)^{\frac{1}{2}}$$
$$= 2\left(1 + (\tfrac{1}{2})\left(\frac{x}{4}\right) + \frac{(\frac{1}{2})(-\frac{1}{2})}{2!}\left(\frac{x}{4}\right)^2 + \ldots\right)$$
$$= 2 + \frac{x}{4} - \frac{x^2}{64} + \ldots \quad |x| < 4$$
$$\therefore \quad f(x) = 1 - 2x + 3x^2 + 2 + \frac{x}{4} - \frac{x^2}{64} + \ldots$$
$$= 3 - 1\tfrac{3}{4}x + 2\tfrac{63}{64}x^2 + \ldots \quad |x| < 1$$

Ignoring terms in x^3 and higher, $f(x) \approx a + bx + cx^2$, with $a = 3$, $b = -1\frac{3}{4}$, $c = 2\frac{63}{64}$.

The restriction is $|x| < 1$.

2 (a)
$$y = \cos^2 x \sin x$$
$$\frac{dy}{dx} = \cos^2 x(\cos x) + \sin x(-2\sin x\cos x)$$
$$= \cos^3 x - 2\cos x\sin^2 x$$
$$= \cos^3 x - 2\cos x(1-\cos^2 x)$$
$$= \cos^3 x - 2\cos x + 2\cos^3 x$$
$$= 3\cos^3 x - 2\cos x$$

(b)
$$u = \cos x \Rightarrow \frac{du}{dx} = -\sin x$$
$$\frac{dx}{du} = -\frac{1}{\sin x} \Rightarrow \sin x\frac{dx}{du} = -1$$
$$\int \cos^2 x \sin x dx$$
$$= \int \cos^2 x \sin x \frac{dx}{du} du$$
$$= \int (-u^2) du$$
$$= -\tfrac{1}{3}u^3 + c$$
$$= -\tfrac{1}{3}\cos^3 x + c$$

(This can be also be done by direct recognition, see page 36.)

3
$$\frac{dy}{dx} = \frac{2x(x+1) - 1(x^2-4)}{(x+1)^2}$$
$$= \frac{x^2+2x+4}{(x+1)^2}$$

When $x = 2$, $\frac{dy}{dx} = \frac{4}{3}$

$\therefore$ gradient of normal $= -\frac{3}{4}$

Equation of normal at (2, 0):
$$y - 0 = -\tfrac{3}{4}(x-2)$$
$$4y = -3x + 6$$
$$3x + 4y - 6 = 0$$

$\therefore$ $a = 3$, $b = 4$ and $c = -6$.

4 (a) $(x-1)^2 + (y-3)^2 = 4$

$CP^2 = (5-1)^2 + (8-3)^2 = 41$

$CA^2 = r^2 = 4$

By Pythagoras' theorem

$AP^2 = CP^2 - CA^2 = 41 - 4 = 37$

By symmetry, $AP = BP = \sqrt{37}$ units.

5
$$x\frac{dy}{dx} = y + yx \Rightarrow x\frac{dy}{dx} = y(1+x)$$

Separating the variables:
$$\frac{1}{y}\frac{dy}{dx} = \frac{1+x}{x}$$
$$\frac{1}{y}\frac{dy}{dx} = \frac{1}{x} + 1$$
$$\int \frac{1}{y} dy = \int \left(\frac{1}{x} + 1\right) dx$$
$$\therefore \quad \ln y = \ln x + x + c$$

When $x = 2$, $y = 4$,
$$\Rightarrow \quad \ln 4 = \ln 2 + 2 + c$$
$$\Rightarrow \quad c = \ln 2 - 2 \quad (\ln 4 = 2\ln 2)$$

So $\ln y = \ln x + x + \ln 2 - 2$
$$\ln\left(\frac{y}{2x}\right) = x - 2$$
$$\frac{y}{2x} = e^{x-2}$$
$$y = 2xe^{x-2}$$

Pure 3 (continued)

6 Let $f(x) = x^3 - 2x^2 + x - 4$
$f(2) = 8 - 8 + 2 - 4 = -2 < 0$

$f(3) = 27 - 18 + 3 - 4 = 8 > 0$

Since $f(2) < 0$ and $f(3) > 0$, there is a value α where $2 < \alpha < 3$, such that $f(\alpha) = 0$.

$f'(x) = 3x^2 - 4x + 1$

Using $x_{n+1} = x_n - \frac{f(x_n)}{f'(x_n)}$ with $x_1 = 2$:

$$x_2 = 2 - \frac{f(2)}{f'(2)} = 2 - \frac{(-2)}{5} = 2.4$$

$$x_3 = 2.4 - \frac{f(2.4)}{f'(2.4)} = 2.4 - \frac{0.704}{8.68} = 2.3188\ ...$$

$$x_3 = 2.319 - \frac{f(2.319)}{f'(2.319)}$$

$$= 2.319 - \frac{0.0345\ ...}{7.857\ ...} = 2.314\ ...$$

$\therefore \quad \alpha = 2.3$ (1 d.p.)

7 (a) $$\int_0^1 xe^{-2x}dx = \left[x(-\tfrac{1}{2}e^{-2x})\right]_0^1 - \int_0^1 (-\tfrac{1}{2}e^{-2x})dx$$

$$= -\tfrac{1}{2}e^{-2} + \tfrac{1}{2}\int_0^1 e^{-2x}dx$$

$$= -\tfrac{1}{2}e^{-2} + \tfrac{1}{2}\left[(-\tfrac{1}{2}e^{-2x})\right]_0^1$$

$$= -\tfrac{1}{2}e^{-2} + \tfrac{1}{2}(-\tfrac{1}{2}e^{-2} - (-\tfrac{1}{2}))$$

$$= -\tfrac{3}{4}e^{-2} + \tfrac{1}{4}$$

(b) (i) $y = xe^{-x}$

$$\frac{dy}{dx} = x(-e^{-x}) + e^{-x} = e^{-x}(1 - x)$$

$\frac{dy}{dx} = 0$ when $x = 1$

When $x = 1$, $y = e^{-1} \Rightarrow A$ is point $(1, e^{-1})$.

(ii) $$V = \pi\int_0^1 y^2dx = \pi\int_0^1 x^2e^{-2x}dx$$

$$= \pi\left(\left[x^2(-\tfrac{1}{2}e^{-2x})\right]_0^1 - \int_0^1 -\tfrac{1}{2}e^{-2x}(2x)dx\right)$$

$$= \pi\left(-\tfrac{1}{2}e^{-2} + \int_0^1 xe^{-2x}dx\right)$$

$= \pi(-\tfrac{1}{2}e^{-2} - \tfrac{3}{4}e^{-2} + \tfrac{1}{4})$ (using part (a))

$= \tfrac{\pi}{4}(1 - 5e^{-2})$

8 $\frac{dP}{dt} \propto P \Rightarrow \frac{dP}{dt} = kP$, where $k > 0$.

Separating the variables

$$\int \frac{1}{P}dP = \int k dt$$

$\Rightarrow \quad \ln P = kt + c$

$\Rightarrow \quad P = e^{kt+c}$

$\Rightarrow \quad P = Ae^{kt}$ (where $A = e^c$)

$t = 0, P = 500 \Rightarrow 500 = A$

$t = 10, P = 1000 \Rightarrow 1000 = 500e^{10k}$

$\therefore \quad 2 = e^{10k}$

$\Rightarrow \quad 10k = \ln 2$

$\Rightarrow \quad k = 0.1 \ln 2$

$\therefore \quad P = 500e^{(0.1 \ln 2)t}$

When $t = 20$,

$P = 500e^{0.1 \ln 2 \times 20} = 500 \times 4 = 2000$

So the population is 2000 when $t = 20$.

9 (a) $$x^3 + xy + y^2 = 7$$

$$3x^2 + x\frac{dy}{dx} + y + 2y\frac{dy}{dx} = 0$$

At (1, 2)

$$3 + \frac{dy}{dx} + 2 + 4\frac{dy}{dx} = 0$$

$$\Rightarrow 5\frac{dy}{dx} + 5 = 0 \text{ i.e. } \frac{dy}{dx} = -1$$

At (1, 2) gradient of normal = 1

Equation of normal is

$y - 2 = 1(x - 1)$

i.e. $y = x + 1$

(b) $x = t^2$, $y = t + \frac{1}{t}$.

$$\frac{dx}{dt} = 2t$$

$$\frac{dy}{dt} = 1 - \frac{1}{t^2} = \frac{t^2 - 1}{t^2}$$

$$\frac{dy}{dx} = \frac{dy}{dt} \times \frac{dt}{dx} = \frac{t^2 - 1}{2t^3}$$

$\frac{dy}{dx} = 0$ when $t^2 - 1 = 0 \Rightarrow t = \pm 1$

When $t = 1$, coordinates are (1, 2).

When $t = -1$, coordinates are (−1, −2).

The stationary points are at (1, 2) and (−1, −2).

Pure 3 (continued)

10 (a) $\sin 2x = 0$ when $2x = 0, \pi, 2\pi, \ldots$
i.e. when $x = 0, \frac{1}{2}\pi, \pi, \ldots$
$\therefore a = \frac{1}{2}\pi$

(b) $$A = \int_0^{\frac{1}{2}\pi} y\,dx = \int_0^{\frac{1}{2}\pi} \sin 2x\,dx$$
$$= -\tfrac{1}{2}\Big[\cos 2x\Big]_0^{\frac{1}{2}\pi}$$
$$= -\tfrac{1}{2}(-1-1)$$
$$= 1$$
Area of R is 1 square unit.

(c) $$V = \pi\int_0^{\frac{1}{2}\pi} y^2\,dx$$
$$= \pi\int_0^{\frac{1}{2}\pi} \sin^2 2x\,dx$$
$$= \tfrac{1}{2}\pi\int_0^{\frac{1}{2}\pi} (1-\cos 4x)\,dx$$
$$= \tfrac{1}{2}\pi\Big[x - \tfrac{1}{4}\sin 4x\Big]_0^{\frac{1}{2}\pi}$$
$$= \tfrac{1}{2}\pi(\tfrac{1}{2}\pi - 0 - 0)$$
$$= \tfrac{1}{4}\pi^2$$
The volume generated is $\frac{1}{4}\pi^2$ cubic units.

11 (a) $$\frac{dy}{dx} = \frac{\sec x\tan x + \sec^2 x}{\sec x + \tan x}$$
$$= \frac{\sec x(\sec x + \tan x)}{\sec x + \tan x}$$
$$= \sec x$$
When $x = \dfrac{\pi}{4}$,
$$\frac{dy}{dx} = \sec\frac{\pi}{4} = \frac{1}{\cos\frac{\pi}{4}} = \frac{1}{1/\sqrt{2}} = \sqrt{2}$$

(b) $$\int_0^{\frac{1}{4}\pi} \tan^2 x\,dx = \int_0^{\frac{1}{4}\pi} (\sec^2 x - 1)\,dx$$
$$= \Big[\tan x - x\Big]_0^{\frac{1}{4}\pi}$$
$$= 1 - \tfrac{1}{4}\pi$$

12 Referring to origin (0, 0)
$\mathbf{a} = 2\mathbf{i} + 2\mathbf{j}$ and $\mathbf{b} = 2\mathbf{i} + 3\mathbf{j} + \mathbf{k}$

Also $\mathbf{b} - \mathbf{a} = \mathbf{j} + \mathbf{k}$

A vector equation for AB is

$\mathbf{r} = \mathbf{a} + \mu(\mathbf{b} - \mathbf{a})$ where μ is a scalar parameter

$\therefore \quad \mathbf{r} = 2\mathbf{i} + 2\mathbf{j} + \mu(\mathbf{j} + \mathbf{k})$

If lines intersect, there are unique values for μ and λ for which

$2\mathbf{i} + 2\mathbf{j} + \mu(\mathbf{j} + \mathbf{k}) = 2\mathbf{j} + \mathbf{k} + \lambda(\mathbf{i} + \mathbf{k})$

Equating coefficients of **i**, **j** and **k**, in turn qives:

$2 = \lambda$

$2 + \mu = 2$

$\mu = 1 + \lambda$

There are no values of μ and λ that satisfy all three equations, so the lines have no common point.

The angle between the lines is the angle between their direction vectors.

$|\mathbf{j} + \mathbf{k}| = \sqrt{2}$ and $|\mathbf{i} + \mathbf{k}| = \sqrt{2}$

$(\mathbf{j} + \mathbf{k}).(\mathbf{i} + \mathbf{k}) = 1$

Let the angle be θ, then

$$\cos\theta = \frac{1}{\sqrt{2}\sqrt{2}} = 0.5$$

$\Rightarrow \quad \theta = 60°$

Statistics 2/3

1 If $\mu = 20$, $\bar{X} \sim N(20, \frac{16}{10})$,

i.e. $\bar{X} \sim N(20, 1.6)$.

Carrying out a two-tailed test at 5% level, reject H_0 if $z < -1.96$ where

$$z = \frac{17.2 - 20}{\sqrt{1.6}} = -2.213...$$

Since $z < -1.96$, reject H_0.

There is evidence that μ is not 20.

2 (a) $p_s = \frac{132}{200} = 0.66$
99% confidence limits

$$= 0.66 \pm 2.576\sqrt{\frac{0.66 \times 0.34}{200}} = 0.66 \pm 0.086$$

99% confidence interval for p

$= (0.66 - 0.086, 0.66 + 0.086)$

$= (0.57, 0.75)$ (2 d.p.)

(b) The probability that the confidence interval does not include p is 0.01.

(c) The critical value of z is smaller, so the width of the confidence interval will be smaller.

3 (a) $f(x) = k$, $2 \leqslant x \leqslant 12$
Total probability = 1

$\Rightarrow 10 \times k = 1$, i.e. $k = \frac{1}{10}$

$\therefore \; f(x) = \frac{1}{10}$, $2 \leqslant x \leqslant 12$

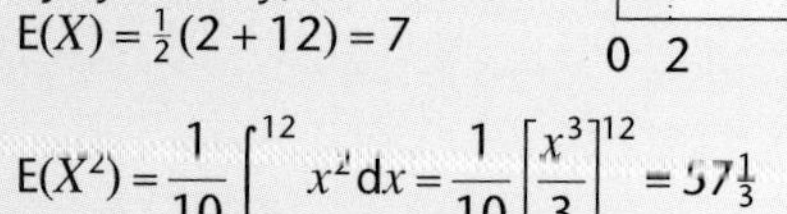

(b) By symmetry,
$E(X) = \frac{1}{2}(2 + 12) = 7$

$$E(X^2) = \frac{1}{10}\int_2^{12} x^2\,dx = \frac{1}{10}\left[\frac{x^3}{3}\right]_2^{12} = 57\tfrac{1}{3}$$

$\text{Var}(X) = 57\frac{1}{3} - 7^2 = 8\frac{1}{3}$

or $\text{Var}(X) = \frac{1}{12}(12 - 2)^2 = 8.333 \ldots$ (formula)

Standard deviation $= \sqrt{8\frac{1}{3}} = 2.887$ (3 d.p.)

$P(7 - 2.887 < X < 7 + 2.887)$

$= P(4.113 < X < 9.887)$

$= 0.1 \times (9.887 - 4.113)$

$= 0.58$ (2 d.p.)

4 For the binomial distribution,
$E(X) = np = 10 \times 0.4 = 4$

$\text{Var}(X) = npq = 4 \times 0.6 = 2.4$

By the central limit theorem, since the sample size is large,

$$\bar{X} \sim N\left(4, \frac{2.4}{60}\right), \text{ i.e. } \bar{X} \sim N(4, 0.04)$$

$$P(\bar{X} < 3.5) = P\left(Z < \frac{3.5 - 4}{\sqrt{0.04}}\right)$$

$$= P(Z < -2.5) = 0.0062$$

5 Let X be the number of trees with the infestation. Assuming that trees are infested independently, with probability p, $X \sim B(n, p)$

$H_0: p = 0.35$

$H_1: p > 0.35$

(a) If $p = 0.35$, $X \sim B(10, 0.35)$
At the 10% level,
H_0 is rejected if $P(X \geqslant 6) < 0.1$.

Using cumulative binomial tables,
$P(X \geqslant 6) = 1 - P(X \leqslant 5)$
$= 1 - 0.9051 = 0.0949$

Since $P(X \geqslant 6) < 0.1$, H_0 is rejected.

There is evidence, at the 10% level, that the percentage of trees that are infested is greater than 35%, indicating that the trees should be felled.

(b) If $p = 0.35$, $X \sim B(30, 0.35)$
n is large such that $np = 10.5 > 5$ and
$nq = 19.5 > 5$ so use normal approximation.
$npq = 10.5 \times 0.65 = 6.825$
so $X \sim N(10.5, 6.825)$ approximately.

At the 10% level, H_0 is rejected if $z > 1.282$.

Test $x = 14$.

Applying a continuity correction,

$$z = \frac{13.5 - 10.5}{\sqrt{6.825}} = 1.148$$

Since $z < 1.282$, H_0 is not rejected.

There is not enough evidence to say that the percentage of infected trees greater than 35%. The trees should be treated with chemicals.

Statistics 2/3 (continued)

6 From the sample: $\bar{x} = \frac{\Sigma x}{n} = \frac{255.6}{9} = 28.4$

(a) $\bar{X} \sim N\left(\mu, \frac{\sigma^2}{n}\right)$, i.e. $\bar{X} \sim N\left(\mu, \frac{4^2}{9}\right)$.

90% confidence limits

$$= \bar{x} \pm 1.645 \frac{\sigma}{\sqrt{n}}$$

$$= 28.4 \pm 1.645 \times \frac{4}{\sqrt{9}}$$

$$= 28.4 \pm 2.193$$

90% confidence interval

$$= (28.4 - 2.193, 28.4 + 2.193)$$

$$= (26.2, 30.6) \text{ (to 1 d.p.)}$$

(b) (i) $\hat{\sigma}^2 = \frac{1}{n-1}\left(\Sigma x^2 - \frac{(\Sigma x)^2}{n}\right)$

$$= \frac{1}{8}\left(7372.26 - \frac{255.6^2}{9}\right) = 14.1525$$

$\hat{\sigma} = 3.762$ (3 d.p.)

(ii) Since the sample size is small, the $t(8)$ distribution is considered. The critical t values, at the 95% level, are ± 2.306.

95% confidence limits

$$= 28.4 \pm 2.306 \times \frac{3.762}{\sqrt{9}} = 28.4 \pm 2.892$$

Interval = (25.5, 31.3) (to 1 d.p.)

7 (a) $F(2) = 1$

$\Rightarrow c(8 - 4 - 1) = 1$ so $c = \frac{1}{3}$

(b) $P(X > 1\frac{1}{2}) = 1 - F(1\frac{1}{2})$

$$= 1 - \tfrac{1}{3}(4 \times 1\tfrac{1}{2} - (1\tfrac{1}{2})^2 - 1)$$

$$= \tfrac{1}{12}$$

(c) For $0 \leqslant x \leqslant 1$,

$$f(x) = \frac{d}{dx}(\tfrac{2}{3}x) = \tfrac{2}{3}$$

For $1 \leqslant x \leqslant 2$,

$$f(x) = \frac{d}{dx}(\tfrac{1}{3}(4x - x^2 - 1))$$

$$= \tfrac{1}{3}(4 - 2x) = \tfrac{2}{3}(2 - x)$$

For $x \leqslant 0$, $x \geqslant 2$, $f(x) = 0$

(d) $E(X) = \int_0^1 \frac{2}{3}x\,dx + \int_1^2 \frac{2}{3}(2x - x^2)dx$

$$= \left[\frac{1}{3}x^2\right]_0^1 + \frac{2}{3}\left[x^2 - \frac{x^3}{3}\right]_1^2$$

$$= \tfrac{1}{3} + \tfrac{2}{3}(4 - \tfrac{8}{3} - (1 - \tfrac{1}{3})) = \tfrac{7}{9}$$

8 H_0: Categories are in the ratio 9 : 3 : 3 : 1
H_1: Categories are not in this ratio.

Expected frequencies:

A $\frac{9}{16} \times 160 = 90$, B $\frac{3}{16} \times 160 = 30$,
C 30 D $160 - 150 = 10$

$\nu = n - 1 = 3$, so consider the $\chi^2(3)$ distribution. From tables, the critical (5%) value is 7.815, so H_0 is rejected if $\chi^2 > 7.815$

	O	E	$\frac{(O-E)^2}{E}$
A	107	90	3.211 ...
B	24	30	1.2
C	23	30	1.633 ...
D	6	10	1.6
	$\Sigma O = 160$	$\Sigma E = 160$	7.644 ...

$$\chi^2 = \sum \frac{(O-E)^2}{E} = 7.644$$

Since $\chi^2 < 7.815$, do not reject H_0.

The results support the theory that the categories are in the ratio 9 : 3 : 3 : 1.

9 Let X be the amount spent in the inner-city store, where $X \sim N(\mu_1, \sigma^2)$.

Let Y be the amount spend in the out-of-town store, where $Y \sim N(\mu_1, \sigma^2)$.

(a) $\hat{\sigma}^2 = \frac{n_1 s_1^2 + n_2 s_2^2}{n_1 + n_2 - 2}$

$$= \frac{60(4.27)^2 + 75(6.89)^2}{60 + 75 - 2} = 34.99 \ldots$$

so $\hat{\sigma} = \sqrt{34.99 \ldots} = 5.916$ (3 d.p.)

H_0: $\mu_1 - \mu_2 = 0$, mean amounts do not differ
H_1: $\mu_1 - \mu_2 \neq 0$, mean amounts differ

At 10% level, H_0 is rejected if $z > 1.645$ or $z < -1.645$.

$$z = \frac{\bar{x} - \bar{y} - 0}{\hat{\sigma}\sqrt{\frac{1}{n_1} + \frac{1}{n_2}}}$$

$$= \frac{19.27 - 21.04}{5.916\sqrt{\frac{1}{60} + \frac{1}{75}}} = -1.727 \ldots$$

Since $z < -1.645$, H_0 is rejected. There is evidence, at the 10% level, that the mean amounts spent at the two stores are different.

Statistics 2/3 (continued)

10 H_0: $\rho_s = 0$ (no correlation)
H_1: $\rho_s < 0$ (disagreement)

From tables, at the 5% level, critical value is −0.6429, so H_0 will be rejected if $r_s < -0.6429$.

	Rank (1)	Rank (2)	$\lvert d\rvert$	d^2
A	3	2	1	1
B	5	1	4	16
C	4	7	3	9
D	2	6	4	16
E	1	8	7	49
F	6	3	3	9
G	7	4	3	9
H	8	5	3	9
				$\Sigma d^2 = 118$

$$r_s = 1 - \frac{6\sum d^2}{n^3 - n} = 1 - \frac{6 \times 118}{512 - 8}$$

$$= -0.4047\ \ldots\ = -0.405 \text{ (3 d.p.)}$$

Since $r_s > -0.6429$, H_0 is not rejected. There is not significant evidence, at the 5% level, of strong disagreement between the rankings of the two judges.

11 Let p be the proportion of voters intending to vote for candidate A.

$p_{s_1} = \frac{236}{500} = 0.472$, $p_{s_2} = \frac{156}{300} = 0.52$;

$$\hat{p} = \frac{n_1 p_{s_1} + n_2 p_{s_2}}{n_1 + n_2} = \frac{236 + 156}{500 + 300} = 0.49$$

H_0: $p_1 = p_2 = p$; H_1: $p_1 \neq p_2$

Performing a two-tailed test, at the 10% level, H_0 is rejected if $z > 1.645$ or $z < -1.645$, i.e. if $|z| > 1.645$.

$$z = \frac{p_{s_1} - p_{s_2}}{\sqrt{\hat{p}\hat{q}\left(\frac{1}{n_1} + \frac{1}{n_2}\right)}}$$

$$= \frac{0.472 - 0.52}{\sqrt{0.49 \times 0.51 \times \left(\frac{1}{500} + \frac{1}{300}\right)}}$$

$$= -1.314$$

Since $|z| < 1.645$, H_0 is not rejected. At the 10% level, there is not a significant difference between the proportions in the two cities intending to vote for candidate A.

12 H_0: Left- or right-handedness and the ability to complete the task are independent.
H_1: There is an association between them.

Expected frequencies

Right-handed, completed task: $\dfrac{150 \times 120}{200} = 90$

	Completed	Not	Totals
Right	90	60	150
Left	30	20	50
Totals	120	80	200

$\nu = (2-1)(2-1) = 1$. Consider the $\chi^2(1)$ distribution. From tables, the critical (5%) value is 3.841, so H_0 is rejected if $\chi^2 > 3.841$.

O	E	$\sum\frac{(\lvert O - E\rvert - 0.5)^2}{E}$
83	90	0.469 …
67	60	0.704 …
37	30	1.408 …
13	20	2.112 …
$\Sigma O = 160$	$\Sigma E = 160$	4.694 …

$$\chi^2 = \sum \frac{(|O - E| - 0.5)^2}{E} = 4.694 \text{ (3 d.p.)}$$

Since $\chi^2 > 3.841$, H_0 is rejected. There is evidence of an association between right- or left-handedness and the ability to complete the task in the given time.

13 (a) If X is number of people with the health problem, then $X \sim B(80, 0.01)$. n is large and p is small, so $X \sim \text{Po}(np)$ i.e. $X \sim \text{Po}(0.8)$ approximately.

$$P(X < 3) = e^{-0.8} + 0.8\,e^{-0.8} + \frac{0.8^2}{2!}e^{-0.8}$$

$$= 0.953 \text{ (3 s.f.)}$$

(b) If X is number of people with the health problem in a sample of size n, $X \sim B(n, 0.01)$.
If $n > 50$, $X \sim P(0.01n)$ approximately.

$P(X \geqslant 1) = 1 - P(X = 0) = 1 - e^{-0.01n}$

So $P(X \geqslant 1) > 0.95 \Rightarrow 1 - e^{-0.01n} > 0.95$
i.e. $e^{-0.01n} < 0.05$

By trial and improvement, $e^{-2.99} = 0.0502 > 0.05$, $e^{-3} = 0.049 < 0.05$, so $0.01n = 3 \Rightarrow n = 300$. (Log theory could be used here.) The minimum number in the sample should be 300.

Mechanics 2

1 (a) Using $y = x\tan\theta - \dfrac{gx^2}{2v^2}(1+\tan^2\theta)$

gives $2.5 = 20\tan\theta - \dfrac{9.8\times 20^2}{2\times 25^2}(1+\tan^2\theta)$

$3.136\tan^2\theta - 20\tan\theta + 5.636 = 0$

$\tan\theta = 6.082\ \ldots$ or $\tan\theta = 0.2954\ \ldots$

$\tan^{-1}(0.2954\ \ldots) = 16.5^0$

so a possible angle of projection is 16.5^0.

(b) Vertically: using $s = ut + \frac{1}{2}at^2$ gives

$-0.5 = 25\sin 16.5^0 t - 4.9t^2$

$4.9t^2 - 25\sin 16.5^0 t - 0.5 = 0$

$t = 1.516\ \ldots$ or $t = -0.067\ \ldots$

since $t > 0$

$d = 25\cos 16.5^0 \times 1.516 - 20$

The distance is 16.3 m to 3 s.f.

2 (a) $x = t^3 - 6t^2 + 3$

$\dot{x} = 3t^2 - 12t = 3t(t-4)$

When $t = 0$, $\dot{x} = 0$ so initially at rest. At rest again when $t = 4$

$4^3 - 6\times 4^2 + 3 = -29$

Max distance in negative direction is 29 m.

(b) The particle is at its initial position when

$t^3 - 6t^2 = 0 \Rightarrow t^2(t-6) = 0$

Particle passes through initial position when $t = 6$.

This gives $\dot{x} = 3\times 6^2 - 12\times 6 = 36\ \text{ms}^{-1}$.

3 (a) $\mathbf{r} = (4t^2+5)\mathbf{i} + 2t^3\mathbf{j}$

$\dot{\mathbf{r}} = 8t\mathbf{i} + 6t^2\mathbf{j}$

When $t = 1$, $\dot{\mathbf{r}} = 8\mathbf{i} + 6\mathbf{j}$ so the speed $= 10\ \text{ms}^{-1}$.

(b) P is moving parallel to

$\mathbf{i} + 3\mathbf{j} \Rightarrow 6t^2 = 24t \Rightarrow 6t(t-4) = 0$

Since $t > 0$ this gives $t = 4$.

4 (a) 3 cm (by symmetry).

(b) $\bar{y} = \dfrac{24\times 2 - 4\times 3}{20} = 1.8$ cm

(c)

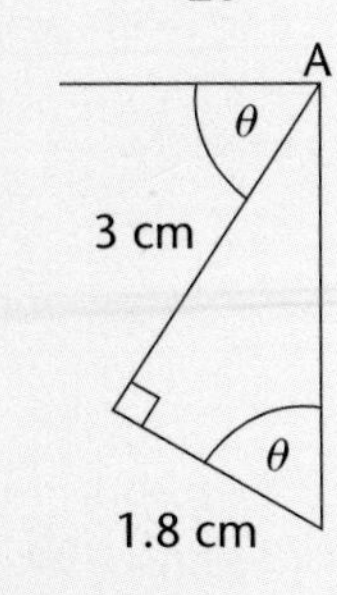

$\theta = \tan^{-1}(\frac{3}{1.8}) = 59.0^0$

5 (a) Area rectangle : area triangle = 3 : 1

so mass of rectangle = 0.6 kg

mass of triangle = 0.2 kg

$\bar{x} = \dfrac{0.6\times 7.5 + 0.2\times 18\frac{1}{3}}{0.8} = 10.2$ cm

(b)

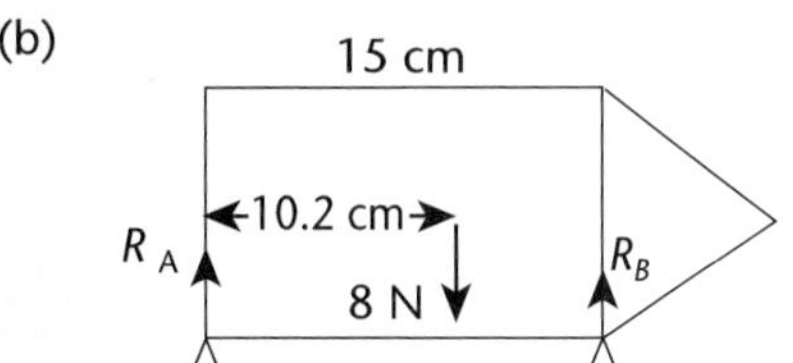

$15R_B = 10.2\times 8 \Rightarrow R_B = 5.44$ N

$R_A = 8 - 5.44 = 2.56$ N

6 (a)

4 ms⁻¹ → ← 2 ms⁻¹

A 0.9   0.3 B

→ u → v

Conservation of momentum:

$0.9\times 4 - 0.3\times 2 = 0.9u + 0.3v$

so $3u + v = 10$ (1)

Newton's experimental law:

$v - u = \frac{1}{3}\times 6 = 2$ (2)

From (1) and (2) $4u = 8$

giving $u = 2$, $v = 4$.

The speeds of A and B after the collision are 2 ms^{-1} and 4 ms^{-1} respectively.

(b) Impulse = change in momentum

$= 0.3\times 4 - 0.3\times(-2)$

$= 1.8$ Ns

(c) Total KE before impact

$= \frac{1}{2}\times 0.9\times 4^2 + \frac{1}{2}\times 0.3\times 2^2 = 7.8$ J

Total KE after impact

$= \frac{1}{2}\times 0.9\times 2^2 + \frac{1}{2}\times 0.3\times 4^2 = 4.2$ J

Loss of mechanical energy = 3.6 J

Mechanics 2 (continued)

7

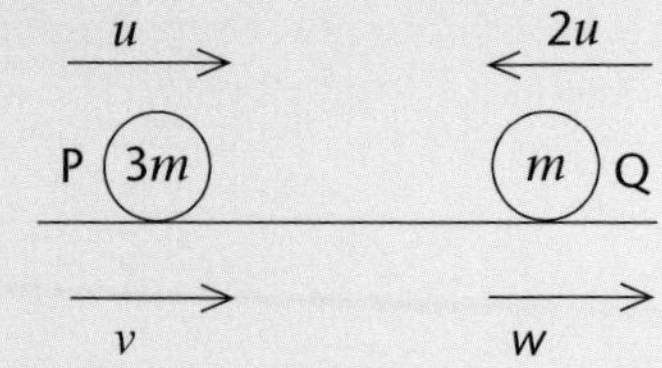

Conservation of momentum:

$3mu - 2mu = 3mv + mw$

giving: $3v + w = u$ (1)

Newton's experimental law:

$w - v = 3ue$ (2)

(1)–(2) gives $4v = u(1 - 3e)$

so $v = \frac{u}{4}(1 - 3e)$

$v < 0 \Rightarrow (1 - 3e) < 0$

$\Rightarrow e > \frac{1}{3}$

8

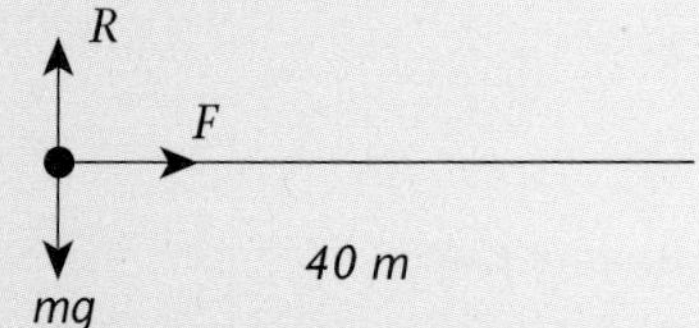

vertically: $R = mg$

horizontally: $F = \frac{m \times 15^2}{40}$

$\mu = \frac{F}{R} = \frac{15^2}{40 \times 9.8} = 0.57$

9

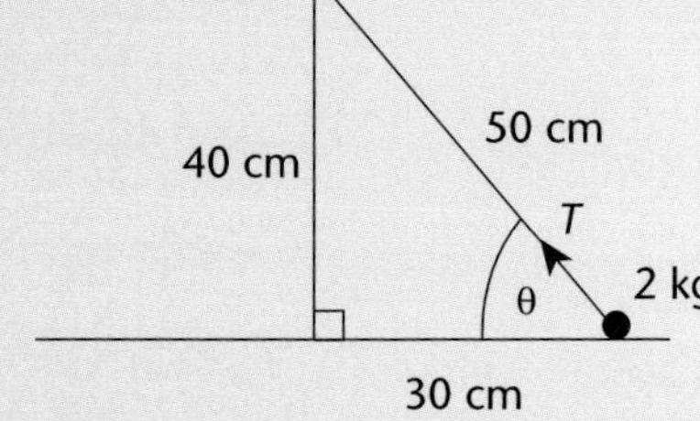

(a) magnitude of acceleration $= \omega^2 r$

$= 3^2 \times 0.3$

$= 2.7 \text{ ms}^{-2}$

(b) Using $F = ma$: $T \cos \theta = 2 \times 2.7$

so: $T \times \frac{3}{5} = 5.4$

giving: $T = 9$ N

(c)

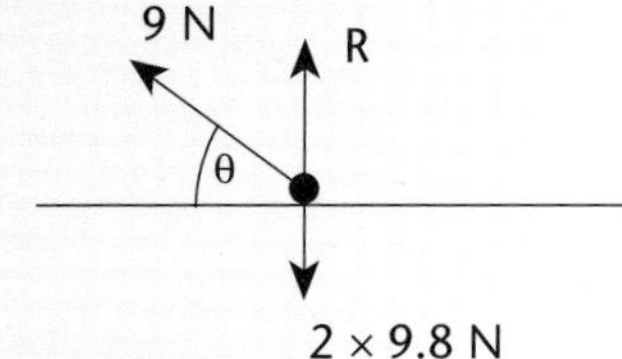

Vertically: $R + 9 \sin \theta = 2 \times 9.8$

$R = 2 \times 9.8 - 9 \times 0.8 = 12.4$

$R = 12.4$ N

10 (a) The tension is perpendicular to the direction of motion.

(b) The acceleration is not constant.

(c) Maximum speed occurs at lowest point.

Loss of PE = gain in KE

$mgl = \frac{1}{2}mv^2$

so, maximum speed is given by $v = \sqrt{2gl}$

11 (a)

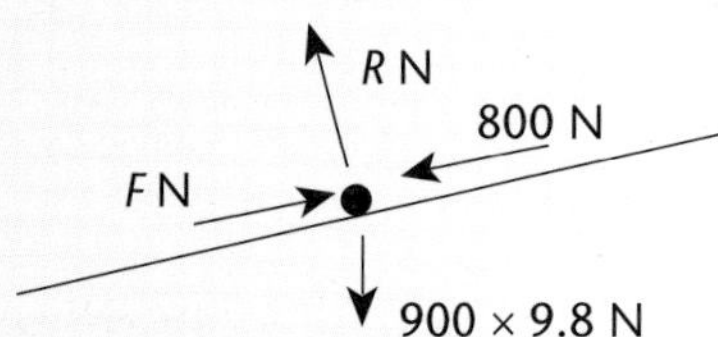

Resolving parallel to the plane:

$F = 900 \times 9.8 \times \frac{1}{15} + 800$

$= 1388$

The driving force is 1388 N

(b) $P = Fv = 1388 \times 20 = 27760$

Power $= 27.76$ kW

(c) Using $F = ma$:

gives: $1388 - 800 = 900a \Rightarrow a = 0.65 \; ms^{-2}$.

12 Momentum of ball before impact $= 0.2 \times 20 = 4$ Ns

Momentum of ball after impact

$= 0.2 \times 5 = 1$ Ns

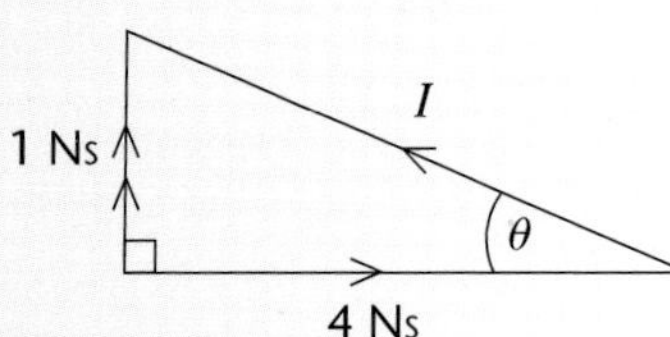

The magnitude of the impulse is given by:

$I^2 = 1^2 + 4^2 = 17 \Rightarrow I = \sqrt{17}$

$\tan \theta = \frac{1}{4} \Rightarrow \theta = 14.0^0$

The impulse has magnitude $\sqrt{17}$ Ns at an angle of 14.0^0 above the horizontal.

Mechanics 3

1 (a) The velocity at time t is given by:
$\dot{x} = 12t^2 - 6t = 6t(2t - 1)$

(b) $\dot{x} = 0$ when $t = 0$ or $t = 0.5$ so the particle changes direction when $t = 0.5$.

$t = 0 \Rightarrow x = 7$

$t = 0.5 \Rightarrow x = 6.75$

The distance travelled is 0.25 m.

(c) P is at its starting point when

$4t^3 - 3t^2 = 0$

$\Rightarrow t^2(4t - 3) = 0$

$\Rightarrow t = 0$, or $t = 0.75$

P returns to its start point after 0.75 s

2 Using $F = ma$: $-kv^2 = m\dfrac{dv}{dt}$

$\dfrac{dv}{dt} = -3$ when $v = 20$

so: $-\dfrac{k}{m} = \dfrac{1}{400} \times -3$

$\Rightarrow \dfrac{dv}{dt} = -\dfrac{3}{400}v^2$

$\Rightarrow \int v^{-2}dv = \int -\dfrac{3}{400}\,dt$

giving: $-\dfrac{1}{v} = -\dfrac{3t}{400} + C$

when $t = 0$, $v = 20$

so: $-\frac{1}{20} = C$

This gives $\dfrac{1}{v} = \dfrac{3t}{400} + \dfrac{1}{20} = \dfrac{3t + 20}{400}$

$\Rightarrow v = \dfrac{400}{3t + 20}$ as required.

When $t = 0$: KE $= \frac{1}{2} \times 2 \times 20^2 = 400$ J

When $t = 4$: KE $= \frac{1}{2} \times 2 \times 12.5^2 = 156.25$ J

Work done by retarding force has magnitude $400 - 156.25 = 243.75$ J.

3 (a) At the lowest point, the loss of gravitational potential energy has been converted to elastic potential energy.

Using $mgh = \dfrac{\lambda x^2}{2l}$

gives $mg(2 + x) = \dfrac{5mgx^2}{4}$

$\Rightarrow 5x^2 - 4x - 8 = 0.$

$\Rightarrow x = 1.73$ or $x = -0.927$ to 3 s.f.

When the particle is brought to rest, the extension is 1.73 m so the object is 3.73 m below O.

(b) Using $T = \dfrac{\lambda}{l}x$

gives $T = \dfrac{5mg}{2} \times 1.73$

Using $F = ma$ gives:

$\dfrac{5mg}{2} \times 1.73 - mg = ma$

Taking $g = 9.8$ gives:

$a = 2.5 \times 9.8 \times 1.73 - 9.8$

$= 32.585$

The object has acceleration 32.6 ms^{-2} to 3 s.f. at its lowest point.

4 (a) Both springs exert an equal force on the particle. The tension in each spring is given by:

$T = \dfrac{\lambda}{l}x = \dfrac{10}{0.8}x$

Using $F = ma$ gives:

$2 \times \dfrac{10}{0.8}x = -0.5\ddot{x}$

$\Rightarrow \ddot{x} = -50x$ as required.

(b) The equation in part (a) shows that the particle moves with SHM where $\omega = \sqrt{50}$.

The periodic time is given by:

$T = \dfrac{2\pi}{\omega} = \dfrac{2\pi}{\sqrt{50}}$

i.e. the time to complete one oscillation is 0.9 seconds to 1 d.p.

(c) The displacement of the particle from O after t seconds is given by:

$x = 0.4\sin(\sqrt{50}t + \alpha)$.

When $t = 0$, $x = 0.4 \Rightarrow \alpha = \dfrac{\pi}{2}$.

In general, $\sin\left(A + \dfrac{\pi}{2}\right) = \cos A$

so $x = 0.4\cos(\sqrt{50}t)$

When $t = 0.3$, $x = -0.209$ … so the distance the particle has travelled is 0.609 … m.

Average speed $= \dfrac{0.609\ \ldots}{0.3} = 2.0$ ms^{-1} to 1 d.p.

Mechanics 3 (continued)

5 (a)

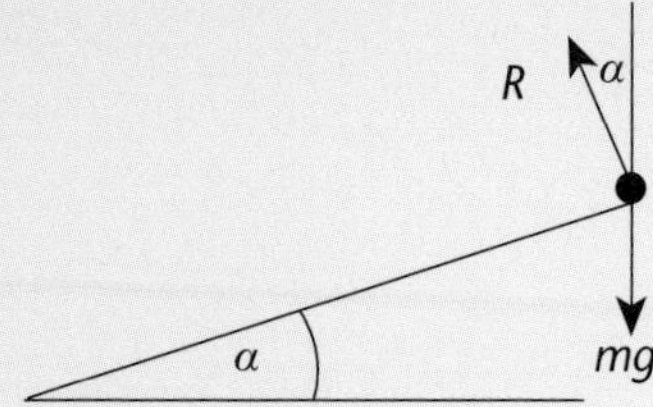

Vertically: $R\cos\alpha = mg$ (1)

Horizontally: $R\sin\alpha = \dfrac{m\times 20^2}{80}$ (2)

(2) ÷ (1) gives $\tan\alpha = \dfrac{20^2}{80\times 9.8}$

$\Rightarrow \alpha = 27.0^0$

(b)

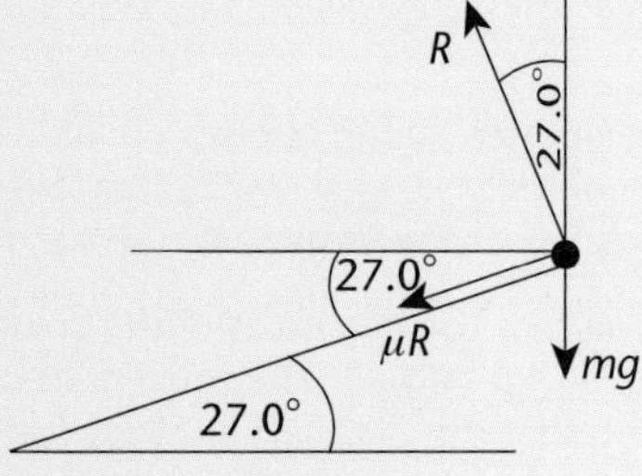

vert: $R\cos 27^0 = mg + \mu R\sin 27^0$ (3)

horiz: $R\sin 27^0 + \mu R\cos 27^0 = \dfrac{m\times 28^2}{80}$ (4)

(4) ÷ (3) gives:

$$\frac{\sin 27^0 + \mu\cos 27^0}{\cos 27^0 - \mu\sin 27^0} = \frac{28^2}{80\times 9.8}$$

rearranging gives:

$$\mu = \frac{28^2\cos 27^0 - 80\times 9.8\sin 27^0}{80\times 9.8\cos 27^0 + 28^2\sin 27^0}$$

$\Rightarrow \mu = 0.325$ to 3 s.f.

6 (a) Horizontally:

$T\sin 30^0 = \dfrac{0.2\times 1.2^2}{0.4}$

$\Rightarrow T = 1.44$

The tension is 1.44 N.

Vertically:

$1.44\cos 30^0 + R = 0.2\times 9.8$

$\Rightarrow R = 0.713$

The reaction is 0.713 N to 3 s.f.

(b) Vertically: the new tension is given by:

$T\cos 30^0 = 0.2\times 9.8 \Rightarrow T = 2.263$

Horizontally: $2.263\sin 30^0 = 0.2\omega^2\times 0.4$

giving: $\omega = 3.76$

The minimum angular speed is 3.76 rad sec^{-1}.

7 (a) By the conservation of energy principle:

KE gained = PE lost

giving: $\frac{1}{2}\times 0.1\times v^2 = 0.1g\times 0.6(1-\cos\theta)$

$\Rightarrow v^2 = 1.2g(1-\cos\theta)$

(b) Resolving along BC:

$0.1g\cos\theta - R = \dfrac{0.1\times 1.2g(1-\cos\theta)}{0.6}$

rearranging gives:

$R = g(0.3\cos\theta - 0.2)$

(c) $R = 0 \Rightarrow 0.3\cos\theta = 0.2$

$\Rightarrow \cos\theta = \frac{2}{3}$

$\Rightarrow \theta = 48.2^0$

8 (a)

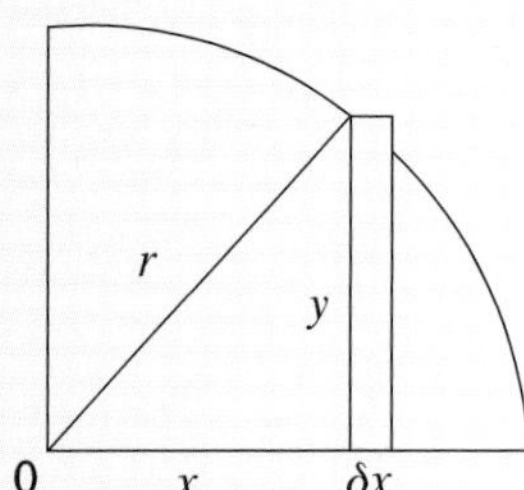

$y = \sqrt{r^2 - x^2}$

Using $\bar{x} = \dfrac{\sum m_i x_i}{\sum m_i}$ and using area to represent mass gives:

$$\bar{x} \approx \frac{\sum x\sqrt{r^2-x^2}\,\delta x}{\dfrac{\pi r^2}{4}}$$

which converts to: $\bar{x} = \dfrac{\int_0^r x\sqrt{r^2-x^2}\,dx}{\dfrac{\pi r^2}{4}}$

(Use integration by substitution to show that the integral is equal to $r^3/3$).

This gives $\bar{x} = \dfrac{\dfrac{r^3}{3}}{\dfrac{\pi r^2}{4}} = \dfrac{4r}{3\pi}$

By symmetry, the coordinates of the centre of mass are $\left(\dfrac{4r}{3\pi}, \dfrac{4r}{3\pi}\right)$.

Mechanics 3 (continued)

(b) Using $\bar{x} = \dfrac{\sum m_i x_i}{\sum m_i}$ and using area to represent mass gives:

$$\bar{x} = \frac{\frac{\pi r^2}{4} \times \frac{4r}{3\pi} + r^2 \times \frac{r}{2}}{\frac{\pi r^2}{4} + r^2}$$

$$= \frac{\frac{r^3}{3} + \frac{r^3}{2}}{\frac{\pi r^2}{4} + r^2}$$

$$= \frac{\frac{r}{3} + \frac{r}{2}}{\frac{\pi}{4} + 1} = \frac{\frac{5r}{6}}{\frac{\pi}{4} + 1}$$

$$= \frac{\frac{10r}{3}}{\pi + 4}$$

so $\bar{x} = \dfrac{10r}{3\pi + 12}$.

Decision and discrete mathematics 2

1 Assume A uses Strategy W with probability p and strategy X with probability $(1 - p)$.

Expected gain for A is:

$4p + 6(1 - p)$ if B uses strategy Y.

$8p + 3(1 - p)$ if B uses strategy Z.

Optimal when

$4p + 6(1 - p) = 8p + 3(1 - p)$

$\Rightarrow -2p + 6 = 5p + 3$

$\Rightarrow p = \frac{3}{7}$

Assume B uses strategy Y with probability q and strategy Z with probability $(1 - q)$.

Expected loss for B is:

$4q + 8(1 - q)$ if A uses strategy W

$6q + 3(1 - q)$ if A uses strategy X.

Optimal when

$4q + 8(1 - q) = 6q + 3(1 - q)$

$\Rightarrow - 4q + 8 = 3q + 3$

$\Rightarrow q = \frac{5}{7}$

The optimal strategy for A is to use strategy W with probability $\frac{3}{7}$ and strategy X with probability $\frac{4}{7}$.

The optimal strategy for B is to use strategy Y with probability $\frac{5}{7}$ and strategy Z with probability $\frac{2}{7}$.

The value of the game is $5\frac{1}{7}$.

2 (a) The minimum row values are –4, –3 and –6. The maximum of these is –3 showing that the play-safe strategy for Richard is strategy B.

The minimum of the negatives of the column values are –4, –6 and –8. The maximum of these is –4 showing that the play-safe strategy for Judy is strategy X.

(b) $-3 + (-4) \neq 0$ so there is no saddle point.

(c) Comparing the row values shows that Richard's pay-offs are *always* less with strategy C than with strategy A. It follows that strategy C will not be a part of his mixed strategy.

(d) Assume Richard uses strategy A with probability p and strategy B with probability $(1 - p)$.

The expected gains for Richard based on the strategy used by Judy are:

X: $4p - 3(1 - p) = 7p - 3$

Y: $3p + 6(1 - p) = -3p + 6$

Z: $-4p + 8(1 - p) = -12p + 8$

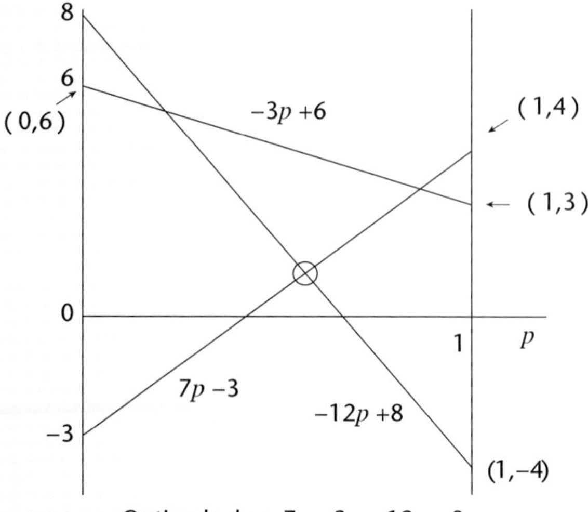

Optimal when $7p - 3 = -12p + 8$

$\Rightarrow p = \frac{11}{19}$

Richard's optimal strategy is to use strategy A with probability $\frac{11}{19}$ and strategy B with probability $\frac{8}{19}$.

Decision and discrete mathematics 2 *(continued)*

3 (a) Using x_n to represent the amount owing at the end of n months gives:

$x_n = x_{n-1} + 0.008x_{n-1} - p$

$\Rightarrow x_n = 1.008x_{n-1} - p$

(b) General solution is:

$$x_n = 1.008^n x_0 - p\frac{(1.008^n - 1)}{0.008}$$

(c) Substituting $x_0 = 15\ 000$ and $x_{60} = 0$ gives:

$$0 = 1.008^{60} \times 15\ 000 - p\frac{(1.008^{60} - 1)}{0.008}$$

$$\Rightarrow p = \frac{1.008^{60} \times 15\ 000 \times 0.008}{1.008^{60} - 1}$$

$= 315.76$ to 2 d.p.

4 (a) $m^2 - 3m - 28 = 0$

(b) $(m-7)(m+4) = 0$

$\Rightarrow m = 7$ or $m = -4$

$x_n = A7^n + B(-4)^n$

(c) Substituting $x_n = an + b$ as a particular solution gives:

$an + b - 3(a(n-1) + b) - 28(a(n-2) + b) = 15n - 36$

$\Rightarrow -30an + 59a - 30b = 15n - 36$

$\Rightarrow a = -0.5$ and $b = \frac{13}{60}$

(d) $x_n = A7^n + B(-4)^n - 0.5n + \frac{13}{60}$

5 (a) 2

(b) $1010 + 0101 = 1111$

$1010 + 1111 = 0101$

$0101 + 1111 = 1010$

Adding any codeword to itself gives 0000 and adding 0000 to any codeword leaves it unchanged.

This shows that for any two codewords belonging to C, their sum belongs to C and so C is a linear code.

(c) There are several possibilities. In this case, you can construct a parity matrix using any two of 1010, 0101 or 1111 to make the rows. (This is not a general result).

6 (a)

$$\begin{pmatrix} 0 & 1 & 0 & 1 & 0 \\ 1 & 0 & 1 & 0 & 0 \\ 1 & 0 & 0 & 0 & 1 \end{pmatrix}\begin{pmatrix} 0 & 1 & 0 & 1 \\ 0 & 0 & 1 & 1 \\ 0 & 1 & 0 & 1 \\ 0 & 0 & 1 & 1 \\ 0 & 1 & 0 & 1 \end{pmatrix} = 0$$

(b)

$$\begin{pmatrix} 0 & 1 & 0 & 1 & 0 \\ 1 & 0 & 1 & 0 & 0 \\ 1 & 0 & 0 & 0 & 1 \end{pmatrix}\begin{pmatrix} 0 \\ 1 \\ 0 \\ 1 \\ 1 \end{pmatrix} = \begin{pmatrix} 0 \\ 0 \\ 1 \end{pmatrix}$$

The error syndrome corresponds to the 5th column of **M**. This suggests that there is an error in the 5th bit of the codeword.

The corrected codeword is 01010.

Index